Sandy Materials in Civil Engineering - Usage and Management

Edited by Saeed Nemati and Farzaneh Tahmoorian

Published in London, United Kingdom

IntechOpen

Supporting open minds since 2005

Sandy Materials in Civil Engineering - Usage and Management
http://dx.doi.org/10.5772/intechopen.78808
Edited by Saeed Nemati and Farzaneh Tahmoorian

Contributors
Yang Jianhui, Nilton De Souza de Souza Campelo, Karine Jussara Sá da Costa, Raimundo Kennedy Vieira, Adalena Kennedy Vieira, Youventharan Duraisamy, David Airey, Gashaw Assefa, Aklilu Gebregziabher, Hebhoub Houria, Kherraf Leila, Abdelouahed Assia, Belachia Mouloud, Parappallil Meeran Rawther Salim, Bellam Siva Rama Krishna Prasad

First published in London, United Kingdom, 2020 by IntechOpen
IntechOpen is the global imprint of INTECHOPEN LIMITED, registered in England and Wales, registration number: 11086078, 7th floor, 10 Lower Thames Street, London, EC3R 6AF, United Kingdom
Printed in Croatia

British Library Cataloguing-in-Publication Data
A catalogue record for this book is available from the British Library

Additional hard and PDF copies can be obtained from orders@intechopen.com

Sandy Materials in Civil Engineering - Usage and Management
Edited by Saeed Nemati and Farzaneh Tahmoorian
p. cm.
Print ISBN 978-1-78985-835-8
Online ISBN 978-1-78985-836-5
eBook (PDF) ISBN 978-1-83962-982-2

For EU product safety concerns:
IN TECH d.o.o., Prolaz Marije Krucifikse Kozulić 3, 51000 Rijeka, Croatia, info@intechopen.com or visit our website at intechopen.com.

Meet the editors

Dr. Saeed Nemati (1969) is a professional writer, translator, editor, journalist, engineer, researcher, and inventor. He is a senior academic member at QIAU and executive director of the World Civil Engineering Information Centre. Dr. Nemati obtained his diploma in building and construction from TAFE Australia. He also received his B.Eng degree in civil engineering, M.Eng degree in environmental engineering, and PhD in infrastructure engineering from Tehran Polytechnic, Tarbiat Modares University (as a top student) and Western Sydney University, respectively. He has about 30 years of work experience in large infrastructure project management. Dr. Nemati is a full member of the Australia Society of Authors with many published ISI articles and books. In addition, he is the winner of 17 national and international awards in civil engineering. His main research interest is innovative ideas in civil engineering with a focus on the design automation and construction robotic technology.

Dr. Farzaneh Tahmoorian works as a lecturer at the Central Queensland University (CQU), Australia, and has broad work experience for over 10 years in many construction projects and projects related to road construction, waste management, landfill construction, etc. She obtained her Ph.D. at Western Sydney University (WSU), Australia; her M.Sc. at the University of Technology, Sydney (UTS), Australia, and her B.Sc. at the International University of Qazvin, Iran. Dr. Tahmoorian is the recipient of various awards such as the Australian Postgraduate Award, Western Sydney Top-up Award, Award of Best Employee of Tehran Municipality, etc. Her research interests are in asphalt mix design, pavement engineering, road and transport engineering, waste management, and sustainability. She is the author of more than 20 books and book chapters and has published in international journals and conference proceedings.

Contents

Preface XI

Section 1
Construction Application of Recycled Sands 1

Chapter 1 3
A Review on the Usage of Recycled Sand in the Construction Industry
by Parappallil Meeran Rawther Salim and Bellam Siva Rama Krishna Prasad

Chapter 2 33
Introduction of Marble Waste Sand in the Composition of Mortar
by Hebhoub Houria, Kherraf Leila, Abdelouahed Assia and Belachia Mouloud

Chapter 3 49
Use of Waste Foundry Sand (WFS) as Filler in Hot-Mixed Asphalt Concrete
by Nilton de Souza Campelo, Karine Jussara Sá da Costa, Raimundo Kennedy Vieira and Adalena Kennedy Vieira

Section 2
Environmental Impacts and Sustainability 67

Chapter 4 69
Environmental Impact and Sustainability of Aggregate Production in Ethiopia
by Gashaw Assefa and Aklilu Gebregziabher

Section 3
Hybrid Aggregates 79

Chapter 5 81
The Influence of Hybrid Aggregates on Different Types of Concrete
by Jianhui Yang

Section 4
Bio-Cemented Sands **107**

Chapter 6 **109**
Geomechanical Behavior of Bio-Cemented Sand for Foundation Works
by Youventharan Duraisamy and David Airey

Preface

As the world moves further into urbanization, there is a greater need for construction materials to meet society's needs.

As natural resources become scarce, the use of recycled materials for construction purposes has become increasingly common. Over the past decade, there has been a significant increase in the utilization of recycled materials in the construction industry. Natural resource depletion, the increasing cost of quarrying, and materials transportation have caused an increase in the cost of natural materials. Therefore, if the use of recycled materials in the construction industry can be justified, it will result in substantial advantages in structure and infrastructure construction coupled with a reduction in the construction cost as well as an improvement in sustainability.

The use of recycled materials in civil projects is not a recent development. In fact, the utilization of recycled materials in construction has a long history. However, finding innovative uses for recycled materials is one of the current priorities in the construction industry.

The aim of compiling this book has been to familiarise the reader with some innovative and alternative construction materials used in civil engineering. The book includes information on using hybrid aggregates, recycled sand, and bio-cemented sands in construction projects.

The editors gratefully acknowledge the contribution of authors and the help of IntechOpen staff in the preparation of this book.

Saeed Nemati
Western Sydney University,
Australia

Farzaneh Tahmoorian
Central Queensland University,
Australia

Section 1

Construction Application of Recycled Sands

Chapter 1

A Review on the Usage of Recycled Sand in the Construction Industry

Parappallil Meeran Rawther Salim and Bellam Siva Rama Krishna Prasad

Abstract

The construction industry requires natural sand for many applications. The recycled sand obtained from industrial operations can also be utilized in construction activities. Many industries generate waste sand as a byproduct. The material generated from this discarded molds and cores is usually known as "used foundry sand," "waste foundry sand," or "spent foundry sand." A vast quantity of used foundry sand is generating and heaping globally in a day-by-day manner. In this review article, an attempt is made to explore the characteristics and various utilizations of this recycled sand from the foundry industry as a valuable resource material in construction activities. From the analysis of the multiple types of research done so far on the reuse of waste foundry sand, it is found that the waste foundry sand is reusable in different civil engineering applications with added advantages. The advantageous applications of used foundry sand are ranging from the road base material to the substitute to fine aggregate in high-performance self-compacting concrete. More generally used foundry sand in the range of 10–30% is best suitable as a partial substitute to regular sand in mortar and concrete making. The use of recycled sand such as waste foundry sand in the construction industry can not only eliminate the problems of waste management and environmental impacts but also substantially boost up the sustainable developmental activities by way of reducing the consumption of natural resources.

Keywords: concrete, mortar, road base, used foundry sand, workability

1. Introduction

From the early civilization of humankind, as a construction material, sand is used extensively. In clay bricks, mud mortars, and lime mortars, the sand finds applications. When the concrete emerged as a construction material, the sand became an integral part of this versatile material. After the stone age, metals were used in making the tools and utensils. For the production of metal items, foundry operations are invariably required. For mold making, the sand has been used from the beginning of the usage of metals in the day-to-day life of the human being. The foundries utilize the sand for two critical applications of mold making and core making. As stated by Javed and Lovell [1], the mold is the outside container of the casting, and the core is the form for achieving the internal shape and cavities within the casted metal structure. Now also, the sand is the primary constituent matter of the mold or core making materials. Silica sand, along with binders such as clay or sodium chemical binders, is used for the making of the mold for metal casting. Sand with clay is naturally available and is abundantly used in mold making. As per the Bureau of Indian Standards IS 3343 [2], original molding sand used in foundries has clay content varied from 5 to 20%. Dogan-Saglamtimur [3], from the research on the reuse of the waste foundry sand in the manufacture of geopolymer concrete, depicted that foundry sands have a loose structure in nature. After several cycles of reuses, the molding sand or foundry sand from the released molds and cores discarded and becomes industrial wastes. A large quantity of such industrial waste sand is generated from foundries all over the world. U. S. Department of Transportation [4] estimated that 1 ton of casting requires 1 ton of foundry sand approximately. American Foundry Society [5] reported that 100 million tons of sand are using and reusing in foundries per annum in the United States itself. The sand from industrial wastes can be recycled for applications in civil engineering and is called "used foundry sand," "waste foundry sand," or "spent foundry sand."

As the quantity of the used foundry sand is so enormous, the disposal of used foundry sand in landfills is significantly affecting the ecology and the environment. Hence, the reuse of waste foundry sand in civil engineering applications other than a land filling is much beneficial to the society from both economical and environmental point of view. The research on the reuse of waste foundry sand in the construction industry is very much crucial as it is the largest consumer of the virgin sand. As a matter of fact, the natural sources of sand are being affected due to the overconsumption and now facing depletion. Every step toward reducing the use of natural sand by adopting waste foundry sand as a full or partial substitute to natural sand will lead to the preservation of the natural resources and the safe disposal of the industrial waste sand to a beneficial application safeguarding the ecology and the environment. In the following paragraphs, the used foundry sand and their properties are discussed in detail. Also, the utilization of used foundry sand recommended by various researchers in a variety of fields related to the construction industry is addressed with sufficient experimental findings. Since concrete plays a significant role in the construction industry, an extensive study of properties of concrete incorporating used foundry sand is also appended for the easy understanding of the applicability of the used foundry sand in concrete making.

2. Used foundry sand

Used foundry sand (UFS), waste foundry sand (WFS), or spent foundry sand (SFS) is obtained from the released waste molds and cores from the foundries. The released molds' and cores' size and shape are different depending on the casting. The discarded molds and cores can be directly used for filling low lying areas. However, there may be a chance of contamination of the water sources due to the chemicals present in the waste foundry sand. To employ the used foundry sand for other civil engineering applications, further processing is required.

It should be noted that the waste foundry sand may contain metal and debris present in the discarded molds. However, as per the reports of the American Foundry Society [5], in most of the foundries, sand reclamation units are employed for the removal of metal particles and debris from the waste foundry sand for advanced applications. Two types of foundry sand are generated from foundries. These are named "green sand" and "chemically bonded sand," depending on the binders used in the production of mold or core. About 90% of the used foundry sand comes under the category of green sand only. Processed used foundry sand as per Salim et al. [6] is shown in **Figure 1**.

DOI: http://dx.doi.org/10.5772/intechopen.92790

Figure 1.
Used foundry sand.

3. Properties of used foundry sand

The used foundry sand has varied physical, chemical, and mechanical properties. In an examination of the characteristics of waste foundry sand and its leachate, Siddique et al. [7] emphasized that the physical and chemical properties of the used foundry sand mostly depend on the industrial segment for which the casting is made. The physical, chemical, and mechanical properties of used foundry sand are discussed in detail below.

3.1 Physical properties

The physical properties of used foundry sand are showing much diversity across the globe. The green sand and chemically bonded sand have different colors. As per the reports of Federal Highway Administration [8], the color of the green sand is gray or black, and the chemically bonded sand has an off-white or medium tan color. Usually, the size of the majority of the particles in the used foundry sand is in the range of 600–150 microns. The U. S. Department of Transportation [4] stated that the used foundry sand has moderately uniform particle size distribution, with just about 85–95% of the particles between 600- and 150-micron sizes and 5–12% of the particles having less than 75-micron sizes. Usually, the used foundry sand consists of subangular to round-shaped particles. The specific gravity of the used foundry sand depends on the properties of the virgin sand and the type of the binders used. Generally, the specific gravity of spent foundry sand has many variations from foundries to foundries. Javed and Lovell [1] stated that the specific gravity of spent foundry sand varies from 2.39 to 2.55. Bulk density of used foundry sand also depends on the properties of virgin sand and the materials used as binders. Naik et al. [9] reported that the bulk density of used foundry sand varies from 1052 to 1554 kg/m^3. The percentage of the mass of water absorbed to the dry mass of the material is water absorption. As per the values of water absorption results from the earlier studies reported by Javed and Lovell [1], American Foundrymen's Society [10], and Johnson [11], the used foundry sand has water absorption of 0.45%. Later, it was revealed that the water absorption values of used foundry sand have much variation from sources to sources. Naik et al. [9] stated that the water absorption of used foundry sand is in the range of 0.38–4.15%. The fineness modulus of sand depends on the grading of the material. The surface moisture content of the sand can reduce the water requirement of the concrete and mortar mix. Most of the researchers did not report the fineness modulus and moisture content of the used foundry sand. However, Seshadri and Salim [12] and Kewal et al. [13] reported that waste foundry sand has a fineness modulus of 2.28 and 2.45, respectively. As per the physical properties stated by Guney et al. [14], the used foundry sand has a moisture content of 3.25%. A comparative graph of the gradation of natural sand and used foundry sand, as reported by Prabhu et al. [15], is shown in **Figure 2**.

3.2 Chemical properties

The chemical properties of used foundry sand depend on the type of binders used in the foundry sand mixture. Johnson [11] reported that the pH of used foundry sand varies from 4 to 8. The used foundry sand consists of different metal oxides. These include SiO_2, Al_2O_3, Fe_2O_3, CaO, MgO, SO_3, Na_2O, K_2O, TiO_2, Mn_2O_3, and SrO. Etxeberria et al. [16] stated that as far as the chemical constituents of used foundry sand were concerned, silicon dioxide constitutes the maximum contribution with 95.10% and the minimum by sulfur trioxide having a contribution of 0.03% of the total mass of used foundry sand. As per the chemical analysis of used foundry sand reported by American Foundrymen's Society [10], the spent foundry sand has a loss on ignition of 5.15%.

3.3 Mechanical properties

The spent foundry sand has excellent mechanical properties at par with the conventional sand. American Foundrymen's Society [10] stated that the spent foundry sand has an angle of internal friction varying from 33° to 40°, and the California Bearing Ratio (CBR) values range from 4 to 20%. As per the reports of the Ministry of Natural Resources [17], the Micro-Deval Abrasion Loss of used foundry sand is less than 2%, and Magnesium sulfate soundness loss varies from 5 to 15%.

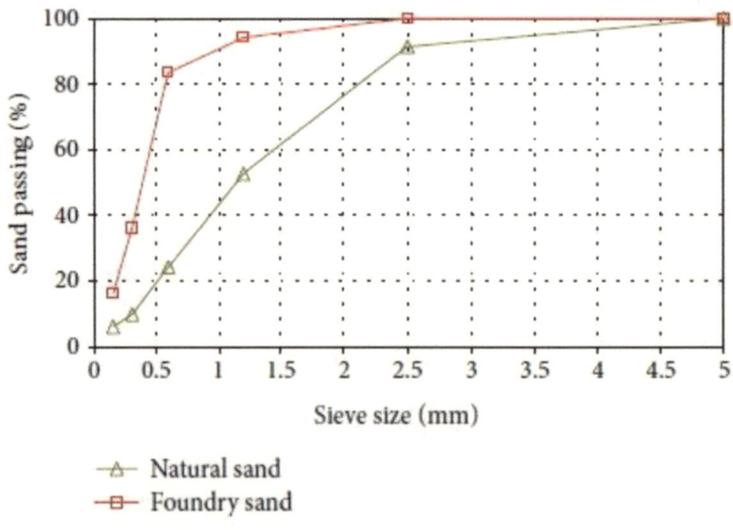

Figure 2.
Gradation of natural sand and foundry sand.

4. Applications of used foundry sand in the construction industry

The used foundry sand can be used in partially or fully for all the purposes where the conventional sand is used. The used foundry sand can be used in a wide variety of applications such as road materials, cement concrete, geopolymer concrete, cement mortars, paver blocks, and masonry blocks.

4.1 Road materials

The used foundry sand can be utilized as materials for road construction. Yazoghli-Marzouk et al. [18] studied the recycling of foundry sand in road construction. They found that treated used foundry sand with a 5.50% hydraulic binder did not show environmental impacts by leaching and has desirable mechanical properties and recommended the application of used foundry sand in the sub-base layer in road construction. The source of foundry sand was a stock of about 150,000 tons of foundry sand stock in Burgundy in France. Iqbal et al. [19] conducted studies on the operation of used foundry sand as a material for embankment, and structural fill further emphasized that sand replaced with 6% used foundry sand is best suitable for structural fill, embankment, and road sub-base material. Generally, it is believed that the compacted waste foundry sand can cause leaching of toxic constituents to the groundwater. But many pieces of research in this regard showed that waste foundry sand did not contaminate the surface water or groundwater. Arulrajah et al. [20] conducted the chemical composition analysis and leachate analysis of used foundry sand. They put forth the implementation of waste foundry sand in road embankment fill and pipe bedding applications. The waste foundry sand used in this research was provided by a recycling plant in Melbourne, Australia. The used foundry sand has superior qualities as that of conventional sub-base material for road construction, and the usage of waste foundry sand can reduce the thickness of the sub-base layer, and thereby, construction cost can be reduced. Guney et al. [21] studied the properties of highway sub-bases with used foundry sand mixtures. They highlighted that the incorporation of used foundry sand can reduce the thickness of the sub-base layer in the sub-base construction of roads.

In the construction of flexible pavements too, the waste foundry sand can be employed to a noticeable extent. The aptness of the waste foundry sand in asphalt mixtures depends on several properties of waste foundry sand, including gradation, particle shape, cleanliness, and surface texture. In a case study on the different methods other than the landfill for the disposal of spent foundry sand generated from the small to medium enterprises in the United Kingdom, Nabhani et al. [22] stated that both green sand and chemically bonded sand could be beneficially replaced with virgin sand in the manufacture of asphalt with an impending extension of its useful working life to about 60 years. Apart from the working life extension, cost savings can also achieve by the replacement of virgin sand by used foundry sand. The used foundry sand incorporated asphalt mixtures are environmentally safe material having no adverse effects on the surroundings. Bakis et al. [23] conducted experiments on the properties of asphalt mixtures made with used foundry sand by replacing the aggregate in different fractions. The environmental impact on the use of used foundry sand also examined. As per the research findings of the investigation on the properties of the used foundry sand incorporated asphalt mixtures, it is described that the use of waste foundry sand in asphalt mixtures did not considerably affect the surrounding environment and further suggested that 10% aggregates can be replaced with the waste foundry sand in the production of asphalt mixtures. Javed et al. [24] investigated the possibilities of the usage of green sand from gray iron castings in asphalt concretes by replacing the total aggregates by 15, 20, and 30% by weight. The bulk-specific gravity, theoretical-specific gravity, Marshall stability, and Marshall flow tests were conducted on the asphalt concrete samples incorporating used foundry sand and control asphalt concrete sample. From the research analysis, it is confirmed that the aggregates in asphalt concrete can be replaced with green sand obtained from gray iron castings up to a replacement level of 15%.

For road foundations also, used foundry sand can be employed in an efficient manner. Pasetto and Baldo [25] investigated the properties of road foundation mixtures made using cement, waste foundry sand, and steel slag in different proportions. The samples were tested after different curing periods for Proctor, compressive strength, indirect tensile strength, and elastic modulus by static and dynamic tests. From the analysis of hydraulically bound mixtures made with waste foundry sand and steel slag, it is noted that the used foundry sand with cement and steel slag showed satisfactory results as per the norms of Italian Road Technical Standards, and the mixture containing 80% of steel slag and 20% of waste foundry sand gives the optimum characteristics. The used foundry sand can be employed in structural fill, embankment, road sub-base, and asphalt concrete mixtures either independently or with other materials like cement and steel slag.

4.2 Cement concrete

The waste foundry sand gradation is mostly outside the lower limits for fine aggregates used in concrete. It is worthwhile to note that the grading of used foundry sand is too fine to satisfy the specifications of fine aggregate. Hence, the waste foundry sand can replace the fine aggregates in cement concrete to some extent only. **Figure 3** shows the grading curve for used foundry sand as per the sieve analysis by Khatib et al. [26] and the gradation limits for fine aggregates as per ASTM-C-33 [27].

4.2.1 General concretes

In general-purpose concretes, used foundry sand finds extensive applications. The used foundry sand is an effective substitute to fine aggregate in general-purpose concretes having strength parameters ranging from low strength to ultra-high strength. Many researchers found that used foundry sand is effective in reducing the usage of fine aggregate in common concretes to a greater extent. Manoharan et al. [28] investigated the characteristics of concrete with chemically bonded used foundry sand in concrete with characteristic compressive strength of 20 MPa having natural river sand replaced with 0, 5, 10, 15, 20, and 25% of used foundry sand and reported that the strength parameters of used foundry sand incorporated concrete containing 5–20% used foundry sand are similar to the control mix with 100% natural river sand of 4.75 mm maximum size as fine aggregate and 20 mm size crushed granite as coarse aggregate. The used foundry sand can reduce the cost of construction, too, to some extent. Bhimani et al. [29] stated that the concrete made with river sand and 20 mm downgraded crushed basalt rock aggregates with a 28th-day compressive strength of 20 MPa and a cost reduction of 3.39% could be achieved by replacing 50% river sand with waste foundry sand in the concrete mix.

In medium strength concrete with a characteristic compressive strength equal to or greater than 30 MPa, used foundry sand is an efficient replacement material to fine aggregates, without compromising on the qualities of the concrete produced. Sohail et al. [30] investigated the properties of concrete of a characteristic compressive strength of 30 MPa made with river sand as fine aggregate and

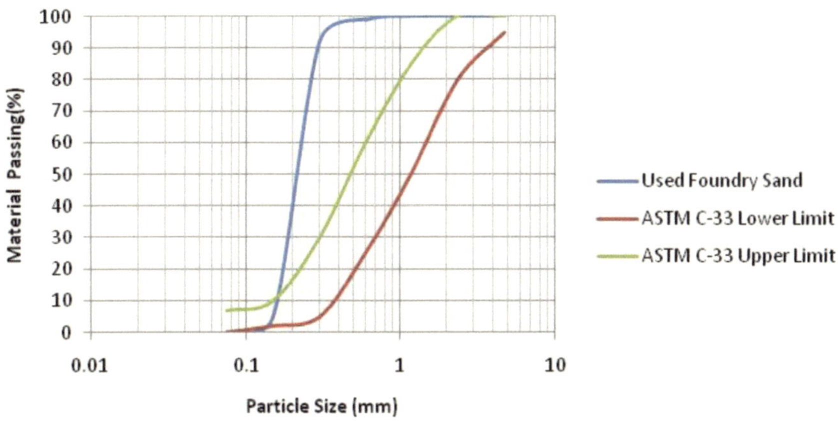

Figure 3.
Used foundry sand grading curve.

20 mm nominal size crushed granite rock aggregates along with green sand from gray iron foundry as a substitute to river sand at 0, 10, 20, 30, 40, 50, 60, 70, 80, 90, and 100% and reported that up to 70% river sand could be replaced with used foundry sand for the concrete with sufficient strength parameters. The abrasion resistance and the strength properties of concrete having 40 MPa compressive strength at 28 days made with 4.75 mm nominal size natural sand and 12.50 mm nominal size coarse aggregate with used foundry sand as a partial substitute to river sand at 0, 5, 10, 15, and 20% were investigated by Singh and Siddique [31] and reported in similar lines that up to 20% natural sand could be replaced with used foundry sand for the production of the concrete having desirable properties and further notified that the incorporation of used foundry sand increased the abrasion resistance of the concrete.

Natural fine aggregates can be replaced with waste foundry sand for the production of high strength concrete also. Guney et al. [14] investigated the application of waste green foundry sand in high strength concrete of compressive strength of 65 MPa at 28 days made with finc sand replaced with waste foundry sand at 0, 5, 10, and 15% by weight of fine sand. They reported that the high strength concrete made with a replacement of 10% of fine aggregates with waste foundry sand exhibited strength parameters at par with the control concrete made with fine sand as fine aggregate. In this research, it is further noted that the freezing and thawing reduced the physical and mechanical properties of concrete by the addition of waste foundry sand to the concrete; however, the strength parameters were found to be acceptable as per the norms fixed by the American Concrete Institute. Chandrasekar et al. [32] succeeded in developing high strength concrete with green sand by partially replacing river sand of 4.75 mm maximum size by 0, 10, 20, 30, and 40% with waste foundry sand and 12.5 mm nominal size coarse aggregate. Slump, compressive strength, split tensile strength, flexural strength, and modulus of elasticity were determined on the samples produced. The effects of the concrete on elevated temperature were also studied. Based on the analysis of the test results, it is confirmed that it is very much possible to replace the fine aggregates with used foundry sand in the range of 10–20% for the production of high strength concrete having a 28th-day compressive strength of 60 MPa for better strength characteristics than the control concrete.

Experimentally, it is proved that ultra-high-strength concrete can be made with used foundry sand as a partial substitute to fine aggregate. Torres et al. [33]

investigated the properties of ultra-high-strength concretes of 120 MPa compressive strength at 28 days made with 3.35 mm well-graded manufactured sand from limestone and river sand as fine aggregates and 6.35 mm size limestone and pea gravel as coarse aggregates along with spent foundry sand. In this research, fine aggregates were replaced by foundry sand at 0, 10, 20, and 30%. As an outcome of the study, it is noted that for optimum performance of the ultra-high-strength concrete, the river sand could be replaced with 10% spent foundry sand in the mix, which uses no coarse aggregates at all.

Nowadays, the requirements of fresh concrete in all the infrastructure projects are met with the ready-mixed concrete (RMC). By using the ready-mixed concrete of required grade, the quality of the concrete can be maintained better than the site mixed concrete. Many kinds of research were conducted on the feasibility of employing used foundry sand in the production of ready-mixed concrete also. Basar and Aksoy [34] conducted experiments on the effect of waste foundry sand as a partial substitute in 0, 10, 20, 30, and 40% of regular sand on the mechanical, leaching, and microstructural characteristics of ready-mixed concrete. The results of the various tests revealed that the typical regular sand in the replacement level of 20% with used foundry sand gives satisfactory mechanical and physical properties in the ready-mixed concrete incorporating used foundry sand.

4.2.2 Special concretes

The used foundry sand can be employed in special concretes like high-performance concrete, self-compacting concrete, high-performance self-compacting concrete, and lightweight concrete. Salim et al. [6] stated that high-performance concrete is high-strength concrete, having desired properties and uniform characteristics. Seshadri and Salim [12] investigated the features of high-performance concrete of design compressive strength of 60 MPa at 28 days with manufactured sand and 20 mm nominal size crushed stone aggregates as fine aggregates and coarse aggregates, respectively, in which fine aggregates were partially replaced by chemically bonded used foundry sand from 0 to 40% in 5% increments and found that up to 30% manufactured sand can be replaced with used foundry sand in the production of high-performance concrete with satisfactory strength characteristics. Ranjitham et al. [35] investigated the properties of 75 MPa characteristic compressive strength high-performance concrete made with 12.5 mm maximum size coarse aggregate and 4.75 mm maximum size river sand as fine aggregate with partial replacement of river sand by green foundry sand and reported that 10% addition of used foundry sand gives excellent strength properties than that of the control concrete without used foundry sand for high-performance concrete of 75 MPa characteristic compressive strength.

For the manufacture of self-compacting concrete also, the used foundry sand can be employed for the reduction in the consumption of the natural fine aggregates. The self-compacting concrete is a type of concrete that does not need external mechanical vibration for the compaction. The self-compacting concrete having strength characteristics in line with the concrete with conventional fine aggregates can be made with partial replacement of fine aggregates with used foundry sand. Siddique and Sandhu [36] reported that self-compacting concrete having a design characteristic compressive strength of 30 MPa made with 15% normal sand replaced by waste foundry sand and 10–12 mm maximum size coarse aggregate exhibited sufficient strength characteristics. Nirmala and Raviraj [37] conducted experiments on the optimization of the self-compacting concrete with used foundry sand as a partial substitute for manufactured sand (M-sand) using the Taguchi approach. The slump flow, V-funnel flow, U-box, L-box, and compressive strength tests were conducted. On the basis of the results obtained, it is noted that for obtaining optimum strength properties for the self-compacting concrete, 20% of manufactured sand (M-sand) should be replaced with spent foundry sand.

In modern construction practice, high-performance self-compacting concrete has great applications where the complicated molds are in use, and the reinforcement steels are very much congested. In this particular situation also, foundry sand waste can be employed with other materials in the production of self-compacting concrete. The high-performance self-compacting concrete has superior early as well as long-term durability and mechanical strength parameters. Makul [38] investigated the properties of high-performance self-consolidating concrete made with waste rice husk ash and foundry sand waste with water to binder ratios of 0.35 and 0.45 where the ordinary portland cement was replaced by rice husk ash in 10 and 20% by weight and the fine aggregate was replaced with foundry waste sand in 30 and 50% by weight. The foundry sand waste used was obtained from automobile part casting foundry. The slump flow, V-funnel flow, splitting tensile strength, and compressive strength tests were performed. Based on the test results, it is observed that the high-performance self-compacting concrete made with 30% replacement of fine aggregates with foundry sand waste and 10% cement replaced with rice husk ash has higher compressive and tensile strength than the conventional self-compacting concrete of the control mix.

Lightweight concrete is concrete, having less density than the regular concrete. In certain applications, regular concrete cannot be entertained due to its higher dead weight. In such situations, lightweight concrete can be effectively utilized. For the manufacture of lightweight concrete also, used foundry sand can be employed efficiently. Hossain and Anwar [39] reported that by the use of waste foundry sand and volcanic ash, lightweight concrete (LWC) can be made economically for the promotion of sustainable construction by reducing the disposal problems of waste foundry sand and volcanic ash.

4.3 Geopolymer concrete

Geopolymer concrete is an innovation in the field of concrete in which cement is not a constituent. In geopolymer concrete also, the waste foundry sand can be used in place of fine aggregates in various replacement levels. Dogan-Saglamtimur [3] investigated the waste foundry sand usage in geopolymer concrete made with sodium hydroxide or sodium silicate for building material production and maximum compressive strength of 12.3 MPa obtained for waste foundry sand incorporated geopolymer concrete containing 30% sodium silicate when the samples were cured at 200°C. The waste foundry sand used in this research is of green sand, which contained bentonite. Based on the results obtained, it is confirmed that the geopolymer material produced with waste foundry sand is suitable for use as a building wall material. For the manufacture of geopolymer concrete cured in ambient temperature also, used foundry sand can be employed in place of fine aggregates. Bhardwaj and Kumar [40] studied the effect of green sand from the ferrous foundry on ambient cured geopolymer concrete. They stated that up to 60% replacement level of fine aggregates to waste foundry sand, the strength parameters are improved better than that of the conventional geopolymer concrete. Scanning electron microscope (SEM) image of concrete of compressive strength of 46 MPa containing 100% chemically bonded foundry sand (FS), as reported by Mavroulidou and Lawrence [41], is shown in **Figure 4**.

In another study on geopolymer concrete made with manufactured sand as fine aggregate with partial replacement of fine aggregate at 0, 5, 10, 15, 20, and 25% by weight of fine aggregate with foundry sand, Jerusha and Mini [42] studied the

Figure 4.
SEM image of concrete containing 100% FS.

slump of the fresh geopolymer concrete and compressive strength of hardened geopolymer concrete samples at 3rd day, 7th day, and 28th day and found that the optimum replacement percentage of foundry sand to the fine aggregate is 15% for the geopolymer concrete made of foundry sand, and the maximum compressive strength obtained was 21.33 MPa.

4.4 Cement mortars

The used foundry sand can constitute as a raw material for the production of cement mortars efficiently. The use of used foundry sand can reduce the cost of the mortars to a considerable extent. Safi et al. [43] conducted experiments on self-compacting mortars made with foundry sand wastes replacing normal sand at 0, 10, 30, and 50% and reported that self-compacting mortars incorporating foundry sand wastes yielded good results at 30% of foundry waste sand in place of normal sand. By the addition of used foundry sand, the workability of the cement mortars gets reduced. However, the deficiency in the workability can be made good by adding a superplasticizer at a low dosage. Cevik et al. [44] investigated the characteristics of cement mortars incorporating waste foundry sand from Turkey steel manufacturer as a partial substitute to natural sand at 0–60%. Based on the compressive strength tests conducted on samples at 3, 7, and 28 days, it is found that the optimum percentage substitution of used foundry sand as a replacement of natural sand in cement mortar is 15%, which yields the maximum compressive strength. Another research study on the use of calcium aluminate cement for recycling green sand and chemically bonded sand conducted by Navarro-Blasco et al. [45] confirmed that by using calcium aluminate cement, mortars of strength higher than 10 MPa can be produced with regular sand replaced by waste foundry sand at 50%.

4.5 Precast concrete products

The used foundry sand can be incorporated in the concrete for the manufacture of precast concrete products like paver blocks and masonry blocks. Many researchers conducted experiments on the applicability of used foundry sand in the production of paver blocks. Marchioni et al. [46] conducted experiments on paver blocks with spent foundry sand in Brazil. They reported that the paver blocks produced with 15% replacement of the fine aggregates with spent foundry sand gave acceptable strength parameters as per Brazilian standards ABNT NBR 9781. The incorporation of used foundry sand has shown a mixed response on the compressive strength of paver blocks. Kewal et al. [13] investigated the properties of paver blocks with geopolymer concrete incorporating used foundry sand and stated that the addition of used foundry decreases the compressive strength of paver blocks made with foundry sand-based geopolymer concrete. In another research on interlocking concrete paving blocks produced with foundry sand waste, Santos et al. [47] conducted compressive strength, measurement of dimension, and water absorption test paver blocks incorporating foundry sand waste. From the results, it is noted that the compressive strength of interlocking paver blocks produced with foundry sand waste is less than the compressive strength of paver blocks produced without foundry sand waste as per the specification laid by the Brazilian standards for the paver blocks. Tausif et al. [48], in a research study on foundry sand use in paver blocks, stated that paver blocks made with 12 mm maximum size coarse aggregate and 4.75 mm maximum size natural sand as fine aggregate with 0.3% synthetic fibers and foundry sand usage at 10% replacement of the fine aggregate showed a maximum compressive strength of 51.48 MPa at 28 days. In another research on the feasibility of used foundry sand in concrete pavers, Kulkarni and Katti [49] studied the properties of concrete pavers made with coarse aggregates of 10 mm maximum size and natural river sand as fine aggregate where the fine aggregates were replaced at 0, 25, 50, 75, and 100% with waste foundry sand from metal casting industries. Water absorption, compressive strength, split tensile strength, flexural strength, and abrasion resistance of the paver blocks were determined. From the test results, it is observed that water absorption increases with the percentage addition of waste foundry sand, whereas the compressive strength, splitting tensile strength, flexural strength, and abrasion resistance of paver blocks incorporating waste foundry sand decrease with the percentage addition of waste foundry sand. However, up to 50% replacements of natural river sand by waste foundry sand, the strength parameters of the paver blocks made are within the acceptable limits set forth by Indian Standard IS 15658 for paver blocks.

The waste foundry sand can be utilized in the production of masonry blocks also. Mahima et al. [50] studied compressive strength, water absorption, block density, drying shrinkage, and moisture movement of high-strength solid masonry blocks utilizing waste foundry sand as a replacement for fine aggregate and stated that at a replacement level of 20–30% of manufactured sand to waste foundry sand, the compressive strength and other parameters of the masonry blocks substantially improved over the regular masonry blocks. In this research, the control mix has a compressive strength of 23.78 MPa, whereas the blocks made with 20% fine aggregate replaced by used foundry sand yielded a compressive strength of 24.53 MPa. Naik et al. [51] studied the properties of concrete products like bricks, blocks, and paving stones incorporating recycled materials like used foundry sand, fly ash, and bottom ash. The brick samples were cast with regular sand, 9.5 mm maximum size crushed limestone chips, fly ash, bottom ash, and used foundry sand at 25 and 35% replacement of regular sand and tested for compressive strength, water absorption, density, and drying shrinkage. The test results confirmed that the concrete bricks with fine aggregates replaced with 25 and 35% ferrous green sand met with the compressive strength requirements as per ASTM C 55 for grade N concrete bricks.

A summary of the research studies described for different applications above is given in **Table 1** for easy reference.

Sl. No.	Researchers	Application of used foundry sand
1	Yazoghli-Marzouk et al. [18]	Sub-base layer in road construction
2	Iqbal et al. [19]	Material for embankment and structural fill
3	Arulrajah et al. [20]	Road embankment fill and pipe bedding
4	Guney et al. [21]	Highway sub-bases
5	Nabhani et al. [22]	Manufacture of asphalt
6	Bakis et al. [23]	Asphalt mixtures
7	Javed et al. [24]	Asphalt concretes
8	Pasetto and Baldo [25]	Road foundation mixtures
9	Manoharan et al. [28]	Concrete of compressive strength of 20 MPa
10	Sohail et al. [30]	Concrete of compressive strength of 30 MPa
11	Singh and Siddique [31]	Concrete of compressive strength of 40 MPa
12	Guney et al. [14]	High-strength concrete of compressive strength of 65 MPa
13	Chandrasekar et al. [32]	High-strength concrete of compressive strength of 60 MPa
14	Torres et al. [33]	Ultra-high-strength concrete of compressive strength of 120 MPa
15	Basar and Aksoy [34]	Ready-mixed concrete
16	Seshadri and Salim [12]	High-performance concrete of design compressive strength of 60 MPa
17	Ranjitham et al. [35]	High-performance concrete of design compressive strength of 75 MPa
18	Siddique and Sandhu [36]	Self-compacting concrete having a compressive strength of 30 MPa
19	Nirmala and Raviraj [37]	Self-compacting concrete
20	Makul [38]	High-performance self-consolidating concrete
21	Hossain and Anwar [39]	Lightweight concrete
22	Dogan-Saglamtimur [3]	Geopolymer concrete
23	Bhardwaj and Kumar [40]	Ambient cured geopolymer concrete
24	Jerusha and Mini [42]	Geopolymer concrete
25	Safi et al. [43]	Self-compacting mortars
26	Cevik et al. [44]	Cement mortars
27	Navarro-Blasco et al. [45]	Cement mortars
28	Marchioni et al. [46]	Paver blocks
29	Kewal et al. [13]	Paver blocks with geopolymer concrete
30	Santos et al. [47]	Paver blocks
31	Tausif et al. [46]	Paver blocks
32	Kulkarni and Katti [49]	Concrete pavers
33	Mahima et al. [50]	High-strength solid masonry blocks
34	Naik et al. [51]	Concrete products like bricks, blocks, and paving stones

Table 1.
Summary of research studies.

5. Properties of fresh concrete made with used foundry sand

The properties of fresh concrete made with used foundry sand vary much to that of standard concrete with regular ingredients. The fresh properties of concrete include the workability, temperature, density, and air content.

5.1 Workability

The workability is an essential parameter of the fresh concrete. In most cases, the workability of the concrete made with used foundry sand decreases as the percentage of used foundry sand increases in the mix. As per Khatib et al. [26], the decrease in workability is attributed to the increase in the fineness of the fine aggregate in the mix. However, some researchers reported equal or slightly higher slump values in concrete made with used foundry sand. Mavroulidou and Lawrence [41] stated that the concrete having 20 MPa compressive strength at 28 days made with 100% chemically bonded waste foundry sand showed 160 mm slump as against 120 mm slump for concrete with regular concrete sand. From the research findings on the use of foundry sand in concrete production, Khatib et al. [26] remarked that the slump dropped approximately in a linear manner from 200 mm for the control mix to zero for the mixes containing 80 and 100% waste foundry sand as the replacement of ordinary sand. Manoharan et al. [28] also confirmed that the slump values of concrete having a design compressive strength of 20 MPa at 28 days made with partial replacement of natural river sand with chemically bonded used foundry sand in 0, 5, 10, 15, 20, and 25% showed a significant decrease in slump value when the used foundry sand content increased in the concrete mix. The same trend was also stated by Bhardwaj and Kumar [52] that the addition of waste foundry sand lowered the workability of geopolymer concrete, and the effect was rapid beyond 40% waste foundry sand replacement level. Some researchers noticed that for concrete incorporating used foundry sand, up to a certain percentage replacement of fine aggregates with used foundry sand, the slump value remains constant. After that, the slump value decreases. In the investigation on the effects of foundry sand as a fine aggregate in concrete production, Prabhu et al. [53] observed that up to 10% replacement of fine aggregate with waste foundry sand, the slump value remains constant as that of the control mix, and after that, the slump values decreased. As per Seshadri and Salim [12], the high-performance concrete of 60 MPa characteristic compressive strength prepared with the fractional replacement of manufactured sand with used foundry sand from 0% to 40% showed a decrease in slump values as the percentage of used foundry sand increased in the concrete mix. In this research, the slump obtained was 140 mm for the control high-performance concrete, and at 40% replacement, the slump value obtained was only 105 mm. Ranjitham et al. [35] observed that for 75 MPa characteristic compressive strength high-performance concrete with cement and fly ash, the slump values consistently reduced from 55 to 42 mm with 0–30% addition of foundry sand. From the research on the effect of used-foundry sand on the mechanical properties of concrete, Siddique et al. [54] stated that the concrete having 28.5 MPa characteristic compressive strength showed a decrease in the slump values when the percentage replacement of used foundry sand increased from 0 to 30%. The concrete mix containing used foundry sand normally requires higher dosages of superplasticizers to maintain the workability. The slump variation of the control mix (CM) and the concrete mix with foundry sand (FS) from 10 to 50% replacement of natural sand when tested immediately after mixing, 30 minutes after mixing, and 60 minutes after mixing as reported by Prabhu et al. [15] for 25 MPa characteristic compressive strength concrete mix is shown in **Figure 5**.

5.2 Temperature

Due to the inclusion of used foundry sand into the concrete mix, the temperature of the fresh concrete mix changes. Much research results are not available in this regard for the temperature variations. The temperature difference of the concrete

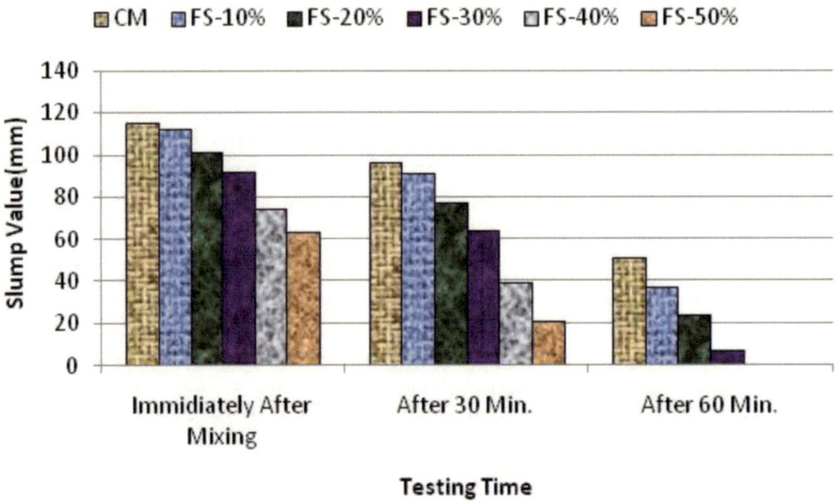

Figure 5.
Workability variation of foundry sand concrete.

mix is attributable to the chemical action of the chemicals present in the used foundry sand with cement and water. In the research report on the application of used foundry sand in concrete production, Prabhu et al. [53] stated that for 20 and 30% replacement of fine aggregate with used foundry sand, the concrete showed an increase in temperature of 1°C from the room temperature. Some researchers observed no variations in fresh concrete temperature to that of room temperature by the addition of used foundry sand in the concrete mix. Manoharan et al. [28] reported that the concrete made with natural river sand replaced with used foundry sand from 10 to 25% in 5% increments had no difference between the room temperature and the fresh concrete temperature. Naik et al. [55] stated that the fresh concrete containing 25 and 35% used foundry sand showed the same temperature as that of room temperature, whereas the control mix showed a 2°C less temperature as that of room temperature. Seshadri and Salim [12] stated that for high-performance concrete made with partial replacement of fine aggregates with used foundry sand, the temperature of the fresh concrete was less than that of the room temperature for all the replacement from 0 to 40%, and the highest temperature difference observed for the replacement of 30 and 35% has a value of 3.5°C, whereas for the control mix, the value observed was 2.5°C.

5.3 Density

The specific gravity of the used foundry sand is normally less than the specific gravity of the fine aggregates. Hence, the density of the concrete incorporating used foundry sand may vary depending on the percentage of the used foundry sand in the concrete mix. Few researchers only reported the density of fresh concrete incorporating used foundry sand. Manoharan et al. [28] stated that the fresh density of concrete made with partial replacement of natural river sand with chemically bonded used foundry sand showed a marginal decrease in fresh density when the used foundry sand content increased from 0 to 25% in the concrete mix, the control mix has a fresh density of 2373 kg/m^3, whereas the concrete containing 25% used foundry sand has a fresh density of 2355 kg/m^3 only.

In some cases, the addition of used foundry sand does not affect the fresh density of concrete. Siddique et al. [54] investigated the effect of used-foundry sand on the mechanical properties of concrete. They reported that the concrete made with used foundry sand showed almost the same fresh density as that of the control mix for 10–30% replacement of fine aggregates with used foundry sand in which the control mix has a fresh density of 2331 kg/m^3, and the concrete with 10, 20, and 30% used foundry sand has a fresh density of 2332 kg/m^3. Naik et al. [55] also observed similar trends and confirmed that the control mix and concrete with 35% used foundry sand have the same fresh density, and fresh density of concrete with 25% used foundry sand has shown an increase in 1.30% over the control mix. For the ultra-high-strength concrete made with used foundry sand, also the fresh density has variation over the control mix. Torres et al. [33] investigated the properties of ultra-high-strength concrete made with silica fume, river sand, steel fibers, and green sand at 0, 10, 20, and 30% by weight of cement. They observed that the fresh density of ultra-high-strength concrete marginally decreased with the increase in the percentage of foundry sand in the mix from 2522 to 2502 kg/m^3.

5.4 Air content

A small quantity of air is entrapped in the concrete. Depending on the concrete mix and type of compaction, the entrapped air content may vary. Manoharan et al. [28] investigated the properties of chemically bonded used foundry sand incorporated concrete and reported that the air content of fresh concrete made with partial replacement of natural river sand with used foundry sand showed a marginal increase with an increase in the used foundry sand content in the concrete mix, the control mix has an air content of 5.2%, whereas the concrete with 25% used foundry sand has an air content of 5.7%. Siddique et al. [54] also observed similar trends in air content for the concrete made with used foundry sand and stated that the air content of the concrete with used foundry sand has a higher percentage of air content than the control mix in which the control mix has an air content of 4.2%, whereas the air content at 10% used foundry sand, the air content value was increased to 4.5%. In some cases, the air content of the concrete mix made with used foundry sand is found to be less than the air content of the regular mix. Naik et al. [55] observed that the air content of the concrete made with used foundry sand tends to decrease up to 25% replacement of fine aggregate with used foundry sand and remains constant further up to 35% replacement.

6. Properties of hardened concrete made with used foundry sand

Many researchers reported the hardened properties of concrete made with used foundry sand at different curing periods. The mechanical properties include compressive strength, split tensile strength, flexural strength, and modulus of elasticity. The mechanical properties of hardened concrete made using waste foundry sand are discussed in detail in the following paragraphs.

6.1 Compressive strength

The concrete incorporating used foundry sand generally shows higher compressive strength than the normal concrete. In some cases, the compressive strength of concrete made with partial replacement of fine aggregates with used foundry sand was below or equal to that of the control mix. Siddique et al. [54] reported that the concrete having the 28th-day compressive strength of 28.5 MPa made with 0, 10, 20, and 30% replacement of sand with used foundry sand, the compressive strength was consecutively increased from 28.5 to 31.3 MPa. Manoharan et al. [28] reported the 28th-day compressive strength of concrete with 0 and 20% chemically bonded used foundry sand as 24.8 and 26.5 MPa, respectively, and for 25% used foundry sand, the compressive strength was below the compressive strength of control mix. In the majority of the research findings, the concrete containing used foundry sand has higher compressive strength than conventional concrete. As per Siddique et al. [54], the increase in compressive strength of the concrete made with used foundry sand may be due to the higher fineness of the used foundry sand than the regular sand, which resulted in the formulation of a denser concrete matrix along with the silica content present in the used foundry sand.

In some cases, the compressive strength of used foundry sand incorporated concrete is more or less the same as that of the control mix up to a certain percentage of used foundry sand content, and after that, the compressive strength decreases significantly. Prabhu et al. [53] stated that the concrete mix containing foundry sand up to 20% replacement of fine aggregate with foundry sand, the compressive strength observed was moderately close to the strength of the control mix, but beyond 20% replacement, the concrete mix showed lower strength than control mix. Some researchers pointed out specific reasons for the reduction of compressive strength of concrete made with used foundry sand beyond certain replacement levels of fine aggregate with used foundry sand. Singh and Siddique [31, 56] and Siddique et al. [57] pointed out that the compressive strength of concrete containing used foundry sand above a particular percentage gets reduced probably due to the increase in surface area of fine particles, which lead to the reduction of water-cement gel in the concrete matrix, and hence, the binding process of the coarse and fine aggregates does not take place properly. The graph of the compressive strength of ultra-high-strength concrete made with natural sand replaced by foundry sand at 0, 10, 20, and 30% at 7, 14, and 28 days as reported by Torres et al. [33] is shown in **Figure 6**.

6.2 Split tensile strength

Depending on the source of used foundry sand, the concrete incorporating used foundry sand shows inferior or at par or superior split tensile strengths than the regular concrete.

In some cases, the split tensile strength of concrete made with used foundry sand increases with the percentage increase in used foundry sand in the concrete mix up to a certain level and decreases afterward. Sohail et al. [30] described that up to 40% replacement of river sand with waste foundry sand from a gray iron foundry, the split tensile strength of concrete at 28th day increases, and further, it reduces consistently up to 100% replacement. Patil et al. [58] confirmed that the split

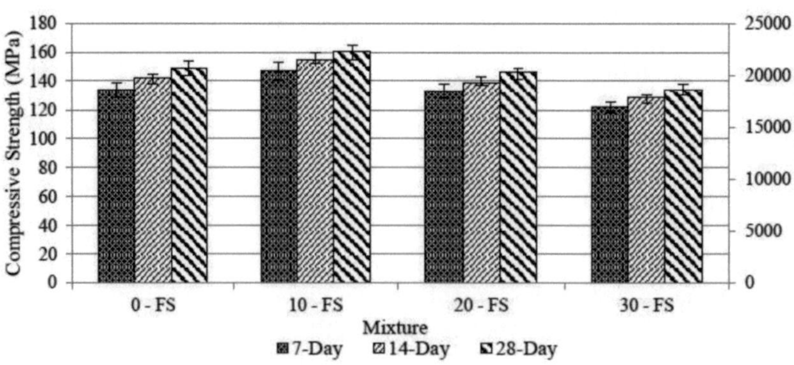

Figure 6.
Compressive strength vs. % foundry sand.

tensile strength of concrete of 30 MPa characteristic compressive strength made with partial replacement of fine aggregates with waste foundry sand increases up to 10% replacement, and further, it decreases in which the control concrete has a split tensile strength of 3.30 MPa, whereas at 10% waste foundry sand content, the split tensile strength increased to 3.87 MPa. Siddique et al. [54] reported that for concrete of 28.5 MPa characteristic compressive strength, the splitting tensile strength was consistently increased from 2.75 to 3.00 MPa from 0 to 30% replacement of regular sand with used foundry sand. In some research findings, the tensile strength of concrete with used foundry sand was found decreasing as the used foundry sand content increases. Seshadri and Salim [12] observed that, for the high-performance concrete with partial replacement of fine aggregate with used foundry sand, the split tensile strength was decreased with the increase in the percentage of used foundry sand from 0 to 40%; at 0% used foundry, the concrete has a split tensile strength of 6.30 MPa, whereas at 40%, the split tensile strength of concrete reduced to 4.40 MPa. Prabhu et al. [53] reported that the split tensile strength of concrete containing foundry sand at 20% substitution of fine aggregate with used foundry sand showed almost equal splitting tensile strength as that of control mix, and the tensile strength in general marginally decreases with an increase in the percentage of foundry sand in the concrete mix. Bhardwaj and Kumar [40] reported that the split tensile strength of ambient cured geopolymer concrete of 40 MPa compressive strength at 28 days made with waste foundry sand increases up to 60% replacement of natural sand with waste foundry sand from the ferrous foundry and decreases afterward for further increase in waste foundry sand percentage. A graphical representation of the split tensile strength of geopolymer concrete of 40 MPa compressive strength at 28 days made of waste foundry sand at different percentage replacements as per Bhardwaj and Kumar [40] is shown in **Figure** 7.

6.3 Flexural strength

The flexural strength of the concrete containing used foundry sand shows marginal variations with the addition of used foundry sand. The flexural strength of concrete incorporating used foundry sand usually increases marginally to that of normal concrete. In the research on the properties of concrete with used foundry

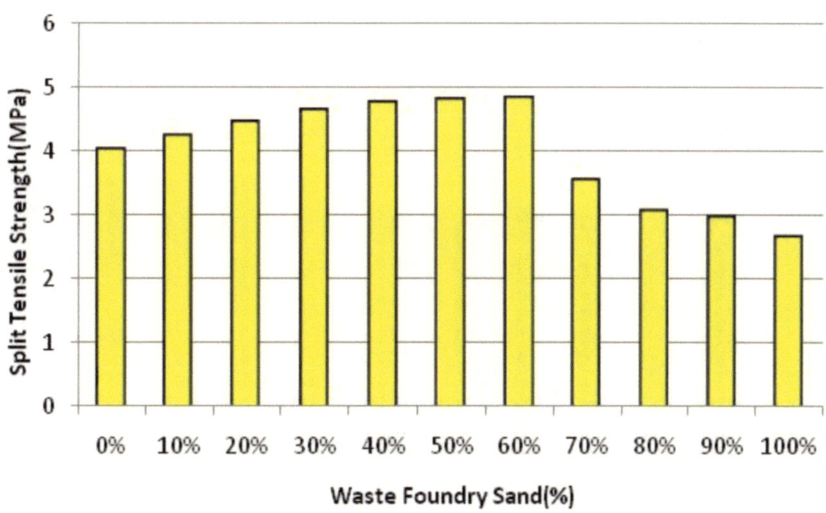

Figure 7.
Split tensile strength vs. % waste foundry sand.

sand, Siddique et al. [54] observed that the flexural strength of 28.5 MPa characteristic compressive strength concrete consecutively increased with the percentage increase of used foundry sand up to 30% replacement level in which the flexural strength of control mix was 3.41 MPa and the flexural strength at 30% replacement was 4.18 MPa.

In some cases, the flexure strength seems to decrease with the increase in percentage addition of used foundry sand. Seshadri and Salim [12] reported that the high-performance concrete having 60 MPa characteristics compressive strength showed a decrease in flexure strength on the increase in replacement of fine aggregate with used foundry sand, in which the control mix has a flexural strength of 10.05 MPa, and at 40% used foundry sand content, the flexural strength decreased to 7.05 MPa. As per Torres et al. [33], at 10% replacement of fine aggregate with foundry sand, the ultra-high-strength concrete showed an increase in flexure strength, and further, it showed a consecutive decrement in flexure strength for 20 and 30% replacement of fine aggregates with foundry sand. Prabhu et al. [53] observed that the flexural strength of concrete with foundry sand content up to 20% of fine aggregates has similar results as that of the control mix; further, the flexural strength decreases after 20% replacement level. The flexural strength variation of concrete having 36.5 and 46 MPa compressive strength at 28 days made with regular sand replaced at 0, 10, 30, 50, 70, and 100% to chemically bonded foundry sand with water to cement ratio 0.55 and 0.45 as reported by Mavroulidou and Lawrence [41] is shown in **Figure 8**.

6.4 Modulus of elasticity

Generally, the modulus elasticity of concrete containing used foundry sand increases up to certain percentage content of used foundry sand and then tends to decrease with further increase in the used foundry sand content. Manoharan et al. [28] observed that the modulus of elasticity of concrete increased with percentage replacement of natural river sand with used foundry sand from 0 to 20%, and further addition of used foundry sand decreased the modulus of elasticity, the modulus of elasticity of control concrete was 23.60 GPa, whereas at 20% replacement of river sand with used foundry sand, the elastic modulus increased to 25.40 GPa. As per the research findings of Prabhu et al. [53], the replacement of fine aggregate

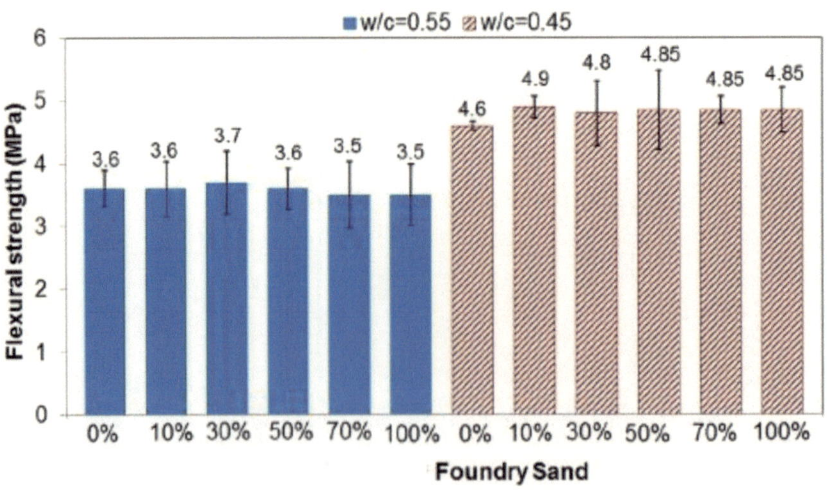

Figure 8.
Flexural strength vs. % foundry sand.

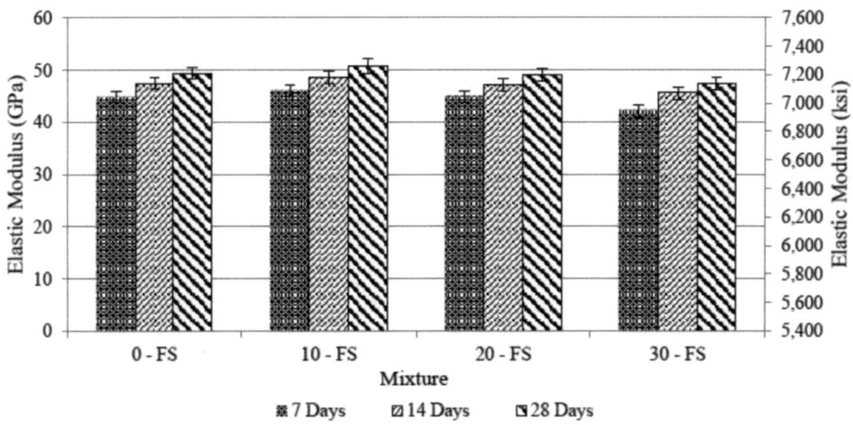

Figure 9.
Elastic modulus variation of foundry sand concrete.

with used foundry sand slightly improved the modulus of elasticity of concrete mix. Some researchers observed marginal reduction of modulus of elasticity by the addition of used foundry sand. Basar and Aksoy [34] stated that the waste foundry sand content in the ready-mixed concrete reduces the modulus of elasticity. The variation of modulus of elasticity of ultra-high-strength concrete made with foundry sand at 7, 14, and 28 days for foundry sand percentages of 0, 10, 20, and 30% as reported in the research findings of Torres et al. [33] is shown in **Figure 9**.

7. Absorption and permeability characteristics of concrete made with used foundry sand

The absorption and permeability characteristics of concrete include water absorption, rapid chloride permeability, sorptivity, and carbonation. The absorption and permeability characteristics of concrete incorporating used foundry sand are discussed in detail in the following paragraphs.

7.1 Water absorption

The concrete made with used foundry sand is generally more permeable than the normal concrete. However, some researchers reported that the inclusion of used foundry sand has no impact on the water absorption of the concrete. The water absorption is somewhat related to the compressive strength also. As per Basar and Aksoy [34], the concrete having higher water absorption has lower strengths. The water absorption of the hardened concrete has a significant effect on the durability characteristics of concrete. Khatib et al. [26] reported that water absorption of the concrete mix containing used foundry sand, the control mix showed the least and increased for 20, 40, 60, 80, and 100% replacement of fine aggregates with foundry sand. It is further confirmed that the water absorption of 56 days cured concrete samples also followed the same trend. Ready-mixed concrete with used foundry sand also showed similar behavior on water absorption. Basar and Aksoy [34] stated that the water absorption of ready-mixed concrete containing waste foundry sand increased with the increase in percentage replacement of fine aggregate with waste foundry sand. Some researchers observed a marginal decrease in water absorption of the concrete containing used foundry sand over the normal concrete. Salokhe and Desai [59] reported that the foundry waste sand had no apparent impact on the water absorption of concrete; however, at 20% ferrous foundry waste sand, the water absorption showed a decrease over the water absorption of the control mix, the control mix has water absorption of 1.91, and at 20% used foundry sand, the water absorption value was 1.13%.

7.2 Rapid chloride permeability

Rapid chloride permeability test (RCPT) is an important test to ascertain the durability of concrete. In this test, as per ASTM C 1202-19 [60], the higher the charge passed through the samples, the concrete is more permeable. The penetration of chlorides through the concrete can affect the reinforcement steel, and the corrosion takes place. Hossain and Anwar [39] studied the rapid chloride penetration of lightweight concrete samples of 20 and 28 MPa compressive strength at 28 days made of waste foundry sand and volcanic ash from Papa New Guinea and reported that the chloride permeability of lightweight concrete decreases with the increase in the percentage content of waste foundry sand. As per the observations of Siddique et al. [57], for 20 and 30 MPa characteristic compressive strength concrete with regular sand partially replaced with spent foundry sand, the charge passed was found to be decreasing with the increase in spent foundry sand content in the concrete mix. In some cases, the chloride permeability decreases up to a certain percentage of used foundry sand in the concrete mix, and further, it increases. Singh and Siddique [31, 56] reported that the chloride permeability of concrete incorporating waste foundry sand decreases up to 15% substitution of fine aggregate with waste foundry sand, and further, it increases. In some cases, the used foundry sand content in the concrete increases the chloride permeability. Aggarwal and Siddique [61] stated that the concrete samples passed charges of 578, 628, 616, 600, 664, 652, and 741 coulombs for 0, 10, 20, 30, 40, 50, and 60% replacement of fine aggregates with waste foundry sand, respectively. As per ASTM C 1202-19 [60], all the above samples have very low permeability as the charges passed were between 100 and 1000 coulombs. A graphical representation of the charges passed through the samples on rapid chloride permeability test (RCPT) at 56 days conducted by Hossain and Anwar [39] on lightweight concrete samples made with waste foundry sand and volcanic ash is shown in **Figure 10**.

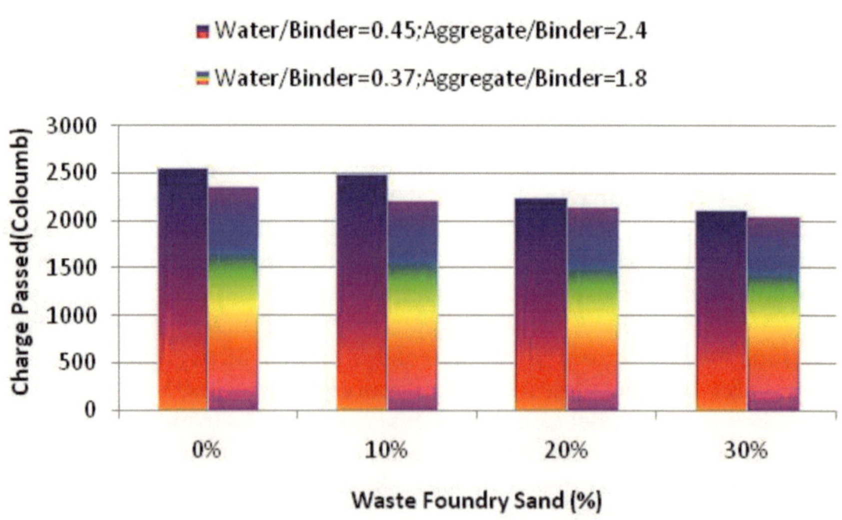

Figure 10.
Chloride penetration of lightweight foundry sand concrete.

7.3 Sorptivity

The sorptivity of the concrete is due to the capillary rise of water from the bottom of the concrete specimen. Some researchers reported a decrease in sorptivity up to certain percentage content of used foundry sand and an increase in sorptivity after that. Bhardwaj and Kumar [40] reported that the sorptivity of geopolymer concrete made with waste foundry sand tends to decrease from 0 to 60% substitution of fine aggregate with waste foundry sand, and further addition of waste foundry sand in the mix increased the sorptivity. It is also observed that for the concrete having up to 80% of waste foundry sand, the initial rate of absorption (IRA) is less than the IRA of the control mix. Khatib et al. [62] reported that for the concrete made with natural sand replaced with used foundry sand at 0, 30, 60, and 100%, waste foundry sand (WFS) exhibited a consecutive increase in water absorption by capillary action when the WFS content increased in the concrete mix. A graph of the sorptivity variation of geopolymer concrete made with waste foundry sand as per Bhardwaj and Kumar [40] is shown in **Figure 11**.

7.4 Carbonation

Carbonation is the reaction of carbon dioxide in the atmosphere with the calcium hydroxide in the cement paste. This reaction produces calcium carbonate and lowers the pH to a value of around 9. The carbonation affects the durability of the concrete. Generally, the used foundry sand content in the concrete mix increases the carbonation depth. Prabhu et al. [15] reported that the carbonation depth on 180 days of the concrete made with used foundry sand increased with the percentage increase in used foundry sand in the concrete mix. At 365 days also, the carbonation depth observed was increasing with the percentage increase in used foundry sand. Siddique et al. [63] stated that the carbonation depth of concrete made with 0, 10, 20, 30, 40, and 50% used foundry sand at 90 and 365 days increased with the used foundry sand content in the mix. The carbonation depth variation at 180th and 365th days as per Prabhu et al. [15] for 25 MPa characteristic compressive strength concrete made with used foundry sand is shown in **Figure 12**.

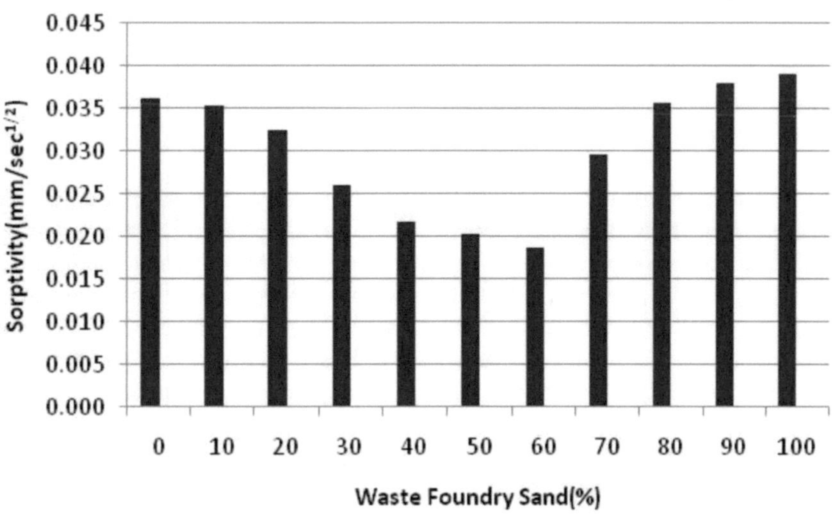

Figure 11.
Sorptivity vs. % waste foundry sand.

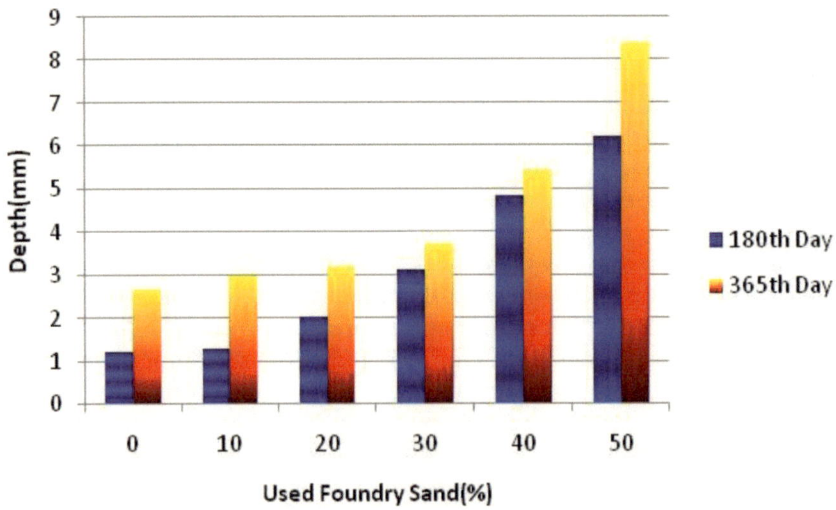

Figure 12.
Carbonation depth vs. % used foundry sand.

8. Ultrasonic pulse velocity (UPV) tests on used foundry sand concrete

The ultrasonic pulse velocity (UPV) test is one of the nondestructive tests (NDTs) to check the quality of the concrete. In this test, the quality and strength of concrete are evaluated by noting down the velocity of an ultrasonic pulse passing through a concrete body. A very few research results are only published on the UPV tests on concrete containing waste foundry sand. Khatib et al. [26] reported that the concrete specimens cured for 28 days showed a consistent decrease in ultrasonic pulse velocity values when the fine aggregates in the concrete mix were replaced with foundry sand in the range of 0, 20, 40, 60, 80, and 100%. The same trend was observed for the specimens cured for 56 days also. Prabhu et al. [15] also stated that the increasing amount of waste foundry sand in the concrete systematically decreases the ultrasonic pulse velocity of concrete made with natural sand replaced with 0, 30, 60, and 100% of waste foundry sand.

9. Long-term strength characteristics of concrete made with used foundry sand

Many research findings are available on the long-term strength characteristics of concrete made with used foundry sand. Siddique et al. [54] studied the long-term strength characteristics of concrete incorporating used foundry sand and reported that the compressive strength, split tensile strength, flexural strength, and modulus of elasticity were improved much at 365 days over the strength at the 28th day for the concrete incorporating used foundry sand. It is to be noted that no detrimental effects were noticed in the strength parameters on aging due to the incorporation of used foundry sand in the concrete mix. Generally, the long-term strength characteristics increase up to certain percentage content of the foundry sand, and the further increase of foundry sand content, the strength decreases. Siddique et al. [63] stated that at 365 days, the compressive strength of concrete increases with percentage replacement of 10, 20, and 30% fine aggregates with foundry sand and decreased for 40, 50, and 60% foundry sand content.

10. Leaching in concrete and mortars made with used foundry sand

The used foundry sand is a nonhazardous material. However, the chemicals present in the used foundry sand can leach into the groundwater and may affect the groundwater quality. As per Siddique et al. [7], the liquid drains or leaches from a landfill are called leachate. The leachate test is essential to assess the suitability of the used foundry sand for certain applications. Very few research observations are available on the leachate analysis of the concrete/mortars made with used foundry sand. Monosi et al. [64] conducted dynamic leaching tests on mortar samples as per Italian standards. They reported that the mortars made from used foundry sand do not release leachate higher than the values specified by Italian standards, and the pH of the leachate was found to be alkaline during the entire testing period. Fero et al. [65] observed that the concentrations of organic compounds in groundwater leached from an iron foundry landfill were below their respective detection limits.

In some cases, the used foundry sand may contain heavy metals. Navarro-Blasco et al. [45] reported that in mortars with used foundry sand, the used foundry sand appeared to be contaminated with heavy metals. In another research conducted by Kaur et al. [66] performed a metal analysis of the leachate obtained from concrete made with untreated and fungal treated waste foundry sand and indicated that waste foundry sand is the contributor of the concentration of leachable metals in concrete containing waste foundry sand. Results from the above research further showed that metal concentration in leachate obtained from fungal treated waste foundry sand incorporated concrete is less than the leachate of untreated waste foundry sand concrete.

11. Conclusion

The foundry industries all over the world generate an enormous quantity of waste sand every year. Many investigations conducted on the reuse of waste foundry sand over the years suggested that the sand discarded from the foundry industries as waste material can be recycled and utilized for beneficial applications in road embankment formation, structural fill, pipe bedding, asphalt concrete, mortars, and different types of concretes. But horizons are still open for the researchers for further innovations in the application of used foundry sand mainly related to the needs in the construction industry where better strength and durability properties are of the paramount concern. In most of the research findings, it suggested that 10–30% fine aggregates can be replaced with used foundry sand for the manufacture of concrete and mortars with sufficient strength parameters with reduced cost. Some researchers estimated that the cost reduction is much significant if the waste foundry sand can be employed in making concrete or concrete products near the foundry industries itself. Due to fine particles present in the used foundry sand, the workability of used foundry sand admixed concrete is profoundly much less than the workability of regular concrete having the same water to binder ratio. However, the researchers suggested that this deficiency can be overcome by adding superplasticizers to the mix. Some researchers pointed out that by performing some inexpensive treatments to the used foundry sand, the strength parameters of used foundry sand incorporated concretes and mortars can be enhanced further. Most of the researchers are in the view that the used foundry sand is a nonhazardous material. However, some researchers suggested that it is better to conduct leachate analysis in advance to avoid the chances of corrosion of the reinforcement if the used foundry sand is proposed to be utilized in the production of concrete for RCC structures. From the analysis of the research works done so far, it can be established that the use of waste foundry sand in the construction industry can not only eliminate the problems of waste management and environmental impacts but also substantially boost up the sustainable developmental activities by way of reducing the consumption of natural resources. However, the feasibility of employing used foundry sand in civil engineering applications in the construction industry will invariably depend on the local cost and the availability of the used foundry sand in the required quantities where the construction work is to be executed. Amidst many research findings and suggestions, the beneficial use of used foundry sand in civil engineering applications is only a bare minimum at present. A collective effort from the researcher community, academicians, and industrialists is highly needed for the full utilization of the recycled used foundry sand from the industrial wastes in the construction industry soon.

Author details

Parappallil Meeran Rawther Salim* and Bellam Siva Rama Krishna Prasad
Civil Engineering Department, GITAM Deemed to be University, Hyderabad, India

*Address all correspondence to: rawther.salim@gmail.com

References

[1] Javed S, Lovell CW. Use of Foundry Sand in Highway Construction. West Lafayette, Indiana: Department of Civil Engineering Purdue University; 1994. DOI: 10.5703/1288284316152

[2] IS:3343. Indian Standard Specification for Natural Moulding Sand for Use in Foundries (IS 3343-1965). New Delhi: Bureau of Indian Standards; 1965

[3] Dogan-Saglamtimur N. Waste foundry sand usage for building material production: A first geopolymer record in material reuse. Advances in Civil Engineering. 2018;**2018**:1-10. DOI: 10.1155/2018/1927135

[4] U.S. Department of Transportation. Federal Highway Administration. User Guidelines for Waste and By-product Materials in Pavement Construction. 1997. Available from: https://www.fhwa.dot.gov/publications/research/infrastructure/structures/97148/fs1.cfm [Accessed: 10 October 2019]

[5] American Foundry Society. The Foundry Industry, Recycling Yesterday, Today & Tomorrow. 2019. Available from: https://afsinc.s3.amazonaws.com/Documents/FIRST/recyclingbrochure_lr.pdf [Accessed: 10 August 2019]

[6] Salim PM, Prasad BSRK, Seshadri ST, Mahima G. A state of the art review on the properties of high performance concrete with used foundry sand and mineral admixtures. International Journal of Civil Engineering and Technology. 2018;**9**(9):1368-1376

[7] Siddique R, Kaur G, Rajor A. Waste foundry sand and its leachate characteristics. Resources, Conservation and Recycling. 2010;**54**(12):1027-1036. DOI: 10.1016/j.resconrec.2010.04.006

[8] Federal Highway Administration. Foundry Sand Facts for Civil Engineers. 2004. Available from: http://www.constructionmidwest.com/msds/tech/FHWA%20Sand%20Specifications.pdf [Accessed: 10 September 2019]

[9] Naik TR, Patel VM, Parikh DM, Tharaniyil MP. Utilization of Used Foundry Sand: Characterization and Product Testing, Report No. CBU-1992-20. Wisconsin: University of Wisconsin-Milwaukee; 1992

[10] American Foundrymen's Society. Alternative Utilization of Foundry Waste Sand. Final Report (Phase I); 1991

[11] Johnson CK. Phenols in foundry waste sand. Modern Casting. 1981;**71**(1):48-49

[12] Seshadri ST, Salim PM. Experimental study on high performance concrete with used foundry sand in partial replacement of fine aggregates. Indian Concrete Journal. 2016;**90**(10):87-94

[13] Kewal, Sharma SK, Gupta H. Development of paver block by using foundry sand based geopolymer concrete. Journal of Today's Ideas–Tomorrow's Technologies. 2015;**3**(2):129-144. DOI: 10.15415/jotitt.2015.32009

[14] Guney Y, Sari YD, Yalsin M, Tuncan A, Donmez S. Re-usage of waste foundry sand in high strength concrete. Waste Management. 2010;**30**(8-9):1705-1713

[15] Prabhu GG, Bang JW, Lee BJ, Hyun JH, Kim YY. Mechanical and durability properties of concrete made with used foundry sand as fine aggregate. Advances in Materials Science and Engineering. 2015;**2015**: 1-11. DOI: 10.1155/2015/161753

[16] Etxeberria M, Pacheco C, Meneses JM, Berridi I. Properties of

concrete using metallurgical industrial by-products as aggregates. Construction and Building Materials. 2010;**24**:1594-1600

[17] Ministry of Natural Resources. Mineral aggregate conservation–Reuse and recycling. In: Report Prepared by John Emery Geotechnical Engineering Limited for Aggregate and Petroleum Resources Section. Peterborough, Ontario: Ontario Ministry of Natural Resources, Queen's Printer for Ontario; 1992

[18] Yazoghli-Marzouk O, Vulcano-greullet N, Cantegrit L, Friteyre L, Jullien A. Recycling foundry sand in road construction–field assessment. Construction and Building Materials. 2014;**61**:69-78. DOI: 10.1016/j.conbuildmat

[19] Iqbal MF, Liu QF, Azim I. Experimental study on the utilization of waste foundry sand as embankment and structural fill. IOP Conference Series: Materials Science and Engineering. 2019;**474**:1-8. DOI: 10.1088/1757-899X/474/1/012042

[20] Arulrajah A, Yaghoubi E, Imteaz M, Horpibulsuk S. Recycled waste foundry sand as a sustainable subgrade fill and pipe-bedding construction material: Engineering and environmental evaluation. Sustainable Cities and Society. 2017;**28**(January):343-349. DOI: 10.1016/j.scs.2016.10.009

[21] Guney Y, Aydilek AH, Demirkan MM. Geoenvironmental behavior of foundry sand amended mixtures for highway subbases. Waste Management. 2006;**26**:932-945. DOI: 10.1016/j.wasman.2005.06.007

[22] Nabhani F, McKei M, Hodgson SNB. A case study on a sustainable alternative to the landfill disposal of spent foundry sand. International Journal of Sustainable Manufacturing. 2013;**3**(1): 1-19. DOI: 10.1504/IJSM.2013.058639

[23] Bakis R, Koyuncu H, Demirbas A. An investigation of waste foundry sand in asphalt concrete mixtures. Waste Management & Research. 2006;**24**(3):269-274. DOI: 10.1177/0734242X06064822

[24] Javed S, Lovell CW, Wood LE. Waste Foundry Sand in Asphalt Concrete, Transportation Research Record, TRB, National Research Council. No. 143727-34. Washington, D.C.; 1994

[25] Pasetto M, Baldo N. Experimental analysis of hydraulically bound mixtures made with waste foundry sand and steel slag. Materials and Structures. 2015;**48**(8):2489-2503. DOI: 10.1617/s11527-014-0333-4

[26] Khatib JM, Baig S, Bougara A, Booth C. Foundry sand utilisation in concrete production. In: Proceedings of the 2nd International Conference on Sustainable Construction Materials and Technologies. Ancona 28-30 June 2010, Italy. Wisconsin, USA: University of Wisconsin Milwaukee Centre for By-products Utilization; 2010

[27] ASTM C33/C33M-18. Standard Specification for Concrete Aggregates. West Conshohocken, PA: ASTM International; 2018. DOI: 10.1520/C0033_C0033M-18

[28] Manoharan T, Laksmanan D, Mylsamy K, Sivakumar P, Sircar A. Engineering properties of concrete with partial utilization of used foundry sand. Waste Management. 2018;**71**:454-460. DOI: 10.1016/j.wasman.2017.10.022

[29] Bhimani DR, Pitroda J, Bhavsar JJ. Used foundry sand: Opportunities for development of eco-friendly low cost concrete. International Journal of Advanced Engineering Technology. 2013;**4**(1):63-66

[30] Sohail M, Wahab A, Khan AM. A study on the mechanical properties of concrete by replacing sand with waste foundry sand. International Journal of

Emerging Technology and Advanced Engineering. 2013;**3**(11):83-88

[31] Singh G, Siddique R. Abrasion resistance and strength properties of concrete containing waste foundry sand (WFS). Construction and Building Materials. 2012a;**28**:421-426. DOI: 10.1016/j.conbuildmat.2011.08.087

[32] Chandrasekar R, Chilabarasan T, Roshan TSA, Visuvasam J. Development of high strength concrete using waste foundry sand. Journal of Chemical and Pharmaceutical Sciences. 2017;**10**(1):348-351

[33] Torres A, Aguayo F, Allena S, Ellis M. Mechanical properties of ultrahigh performance fiber reinforced concrete made with foundry sand. Journal of Civil Engineering and Construction. 2019;**8**(4):157-167. DOI: 10.32732/jcec.2019.8.4.157

[34] Basar HM, Aksoy ND. The effect of waste foundry sand (WFS) as partial replacement of sand on the mechanical, leaching, and micro-structural characteristics of ready-mixed concrete. Construction and Building Materials. 2012;**35**:508-515. DOI: 10.1016/j.conbuildmat.2012.04.078

[35] Ranjitham M, Piranesh B, Vennila A. Experimental investigation on high performance concrete with partial replacement of fine aggregate by foundry sand with cement by mineral admixtures. International Journal of Advanced Structures and Geotechnical Engineering. 2014;**3**(1):28-33

[36] Siddique R, Sandhu RK. Properties of self-compacting concrete incorporating waste foundry sand. Leonardo Journal of Sciences. 2013;**23**(July-December):105-124

[37] Nirmala DB, Raviraj S. Experimental study of optimal self-compacting concrete with spent foundry sand as partial replacement for M-sand using Taguchi approach. SSP–Journal of Civil Engineering. 2016;**11**(1):119-130. DOI: 10.1515/sspjce-2016-0013

[38] Makul N. Combined use of untreated-waste rice husk ash and foundry sand waste in high-performance self-consolidating concrete. Results in Materials. 2019;**1**(August):1-11. DOI: 10.1016/j.rinma.2019.100014

[39] Hossain KMA, Anwar MS. Influence of foundry sand and natural pozzolans on the mechanical, durability, and microstructural properties of lightweight concrete. British Journal of Applied Science & Technology. 2015;**10**(4):1-12

[40] Bhardwaj B, Kumar P. Effect of waste foundry sand addition on strength, permeability, and microstructure of ambient cured geopolymer concrete. IOP Conference Series: Materials Science and Engineering. 2018;**431**:1-8. DOI: 10.1088/1757-899X/431/9/092009

[41] Mavroulidou M, Lawrence D. Can waste foundry sand fully replace structural concrete sand? Journal of Material Cycles and Waste Management. 2019;**21**(3):594-605. DOI: 10.1007/s10163-018-00821-1

[42] Jerusha SJ, Mini M. Experimental study on geopolymer concrete with partial replacement of fine aggregate with used foundry sand. International Journal of Advanced Technology in Engineering and Science. 2015;**3**(1):559-569

[43] Safi B, Sebki G, Chahour K, Belaid A. Recycling of foundry sand wastes in self-compacting mortars: Use as fine aggregates. Sofia: Surveying Geology & Mining Ecology Management. 2017;**17**:177-184. DOI: 10.5593/sgem2017/41

[44] Cevik S, Mutuk T, Oktay BM, Demirbas AK. Mechanical and

microstructural characterization of cement mortars prepared by waste foundry sand (WFS). Journal of the Australian Ceramic Society. 2017;**53**(2):829-837. DOI: 10.1007/s41779-017-0096-9

[45] Navarro-Blasco I, Fernández JM, Duran A, Sirera R, Álvarez JI. A novel use of calcium aluminate cements for recycling waste foundry sand (WFS). Construction and Building Materials. 2013;**48**:218-228. DOI: 10.1016/j.conbuildmat.2013.06.071

[46] Marchioni ML, Lyra J, Pillegi R, Pereira RL, Oliveira C. Foundry sand for manufacturing paving units. In: Proceedings of the 10th International Conference on Concrete Block Paving. 24-26 November, 2012. Shanghai, Peoples Republic of China. Beiging: Small Element Pavement Technologists; 2012

[47] Santos CC, Valentina LOV, Cuzinsky FC, Witsmiszyn LC. Interlocking concrete paving blocks produced with foundry sand waste. Materials Science Forum. 2018;**912**:191-195

[48] Tausif K, Tanmay P, Hussain N, Fenil P, Dhruvang P. Experimental study on use of foundry sand in paver blocks. International Journal of New Technologies in Science and Engineering. 2018;**5**(8):58-64

[49] Kulkarni S, Katti V. Experimental study to determine the feasibility of replacing natural river sand by waste foundry sand in concrete pavers. International Journal of Civil Engineering and Technology. 2017;**8**(9):498-505

[50] Mahima G, Sreevidya V, Salim PM. Waste foundry sand as a replacement for fine aggregate in high strength solid masonry blocks. International Journal of Innovative Research in Science, Engineering and Technology. 2016;**5**(5):6878-6886. DOI: 10.15680/IJIRSET.2016.0505037

[51] Naik TR, Kraus RN, Chun YM, Ramme BW, Singh SS. Properties of field manufactured cast-concrete products utilizing recycled materials. Journal of Materials in Civil Engineering. 2003;**15**(4):400-407

[52] Bhardwaj B, Kumar P. Comparative study of geopolymer and alkali activated slag concrete comprising waste foundry sand. Construction and Building Materials. 2019;**209**:555-565. DOI: 10.1016/j.conbuildmat.2019.03.107

[53] Prabhu GG, Hyun JH, Kim YY. Effects of foundry sand as a fine aggregate in concrete production. Construction and Building Materials. 2014;**70**:514-521. DOI: 10.1016/j.ijsbe.2016.04.006

[54] Siddique R, Schutter GD, Noumowe A. Effect of used-foundry sand on the mechanical properties of concrete. Construction and Building Materials. 2009;**23**(2):976-980. DOI: 10.1016/j.conbuildmat.2008.05.005

[55] Naik TR, Patel VM, Parikh DM, Tharaniyil MP. Utilization of used foundry sand in concrete. Journal of Materials in Civil Engineering. 1994;**6**(2):254-263. DOI: 10.1061/(ASCE)0899-1561(1994)6:2(254)

[56] Singh G, Siddique R. Effect of waste foundry sand (WFS) as partial replacement of sand on the strength, ultrasonic pulse velocity, and permeability of concrete. Construction and Building Materials. 2012b;**26**(1):416-422. DOI: 10.1016/j.conbuildmat.2011.06.041

[57] Siddique R, Singh G, Belarbi R, Mokhtar KA, Kunal. Comparative investigation on the influence of spent foundry sand as partial replacement of fine aggregates on the properties of two grades of concrete. Construction and

Building Materials. 2015;**83**:216-222. DOI: 10.1016/j.conbuildmat.2015.03.011

[58] Patil RN, Mehetre PR, Phalak KT. Cement concrete properties incorporating waste foundry sand. International Journal on Emerging Trends in Technology. 2015;**2**(1): 221-225

[59] Salokhe EP, Desai DB. Application of foundry waste sand in manufacture of concrete. IOSR Journal of Mechanical and Civil Engineering. 2013;**1**:43-48

[60] ASTM C 1202-19. Standard Test Method for Electrical Indication of Concrete's Ability to Resist Chloride Ion Penetration. West Conshohocken, PA, United States: ASTM International; 2019. DOI: 10.1520/C1202-19

[61] Aggarwal Y, Siddique R. Microstructure and properties of concrete using bottom ash and waste foundry sand as partial replacement of fine aggregates. Construction and Building Materials. 2014;**54**:210-223. DOI: 10.1016/j.conbuildmat.2013.12.051

[62] Khatib JM, Herki BA, Kenai S. Capillarity of concrete incorporating waste foundry sand. Construction and Building Materials. 2013;**47**:867-871. DOI: 10.1016/j.conbuildmat.2013.05.013

[63] Siddique R, Aggarwal Y, Aggarwal P, El-Hadj K, Bennacer R. Strength, durability, and micro-structural properties of concrete made with used-foundry sand (UFS). Construction and Building Materials. 2011;**25**(4):1916-1925. DOI: 10.1016/j.conbuildmat.2010.11.065

[64] Monosi S, Tittarelli F, Giosue C, Ruello ML. Effect of two different sources and washing treatment on the properties of UFS by-products for mortar and concrete production. Construction and Building Materials. 2013;**44**:260-266. DOI: 10.1016/j.conbuildmat.2013.02.029

[65] Fero RL, Ham RK, Boyle WC. An Investigation of Groundwater Contamination by Organic Compounds Leached from Iron Foundry Solid Wastes. Des Plaines, Illinois, USA: Final Report to American Foundrymen's Society; 1986

[66] Kaur G, Siddique R, Rajor A. Micro-structural and metal leachate analysis of concrete made with fungal treated waste foundry sand. Construction and Building Materials. 2013;**38**:94-100. DOI: 10.1016/j.conbuildmat.2012.07.112

Chapter 2

Introduction of Marble Waste Sand in the Composition of Mortar

Hebhoub Houria, Kherraf Leila, Abdelouahed Assia and Belachia Mouloud

Abstract

The aim of this research is to study the possibility of the valorization of sand marble waste in mortars as substitute in sand. To achieve this study, sand marble waste is used with weight ratios of 5, 10, 15 and 20% to formulate a mortar with sand marble waste and a control mortar with 0% of sand marble waste. The properties in the fresh state, the mechanical strength, absorption by immersion, and the weight loss as well as the shrinkage and acid attack of each mixture were carried out through the conducted experiments. The different results show that the introduction of recycled sand in the mortars gives good results and it can be used as aggregates.

Keywords: mortar, valorization, sand, waste, marble, substitution, performance, durability

1. Introduction

Aggregates production is mainly ensured by extraction in quarries. There are also two other resources that allow manufacturers to produce aggregates. Then, we can speak of recycled aggregates and artificial aggregates. Recycled materials can be represented an excellent source of aggregates, the use of these materials in concrete would help to the recovery of recycled materials, and also, the protection of natural resources.

This work aims to study the effect of partial substitution of ordinary sand (50% dune sand and 50% sea sand) by marble waste sand from the quarry of Filfila located in Skikda (East Algeria) with rates 5, 10, 15 and 20% on the properties of mortar.

Located 25 km by road, east of Skikda—Algeria, the Djebel Filfila marble deposit was exploited by the company ENAMARBRE from Roman antiquity, and may even be before. This deposit consists of outcrops of marble which can reach a length of 1100 m for a width which varies from 100 to 300 m. The exploitable part of the deposit is composed by a lenticular body which extends over 550 m in length occupying an area of 13 Ha. The maximum depth of the deposit is 180 m which shows that this deposit consisting of marble levels of different colors tends to develop in depth.

The marble of Filfila is made up of the following varieties: white, with gray, light gray, dark gray, banded in green shade, banded in brown shade. White marbles represent 44% of the whole.

The Filfila deposit is divided into two quarries:

- The first one is a quarry with white marble blocks and reseda green, the exploitation is carried out by horizontal sawing methods by cutting vertical and lateral sawing by a diamond wire with cooling in clear water. The waste from this quarry is the scraps and rubble having different geometric shapes (**Figure 1**) and declassified powder (powder subject to weathering), the waste rate is 56% of production. The waste from the processing plant is the fall of block sizes and the fall of tiles and marble powders, and the waste rate is 22 m^2/m^3.

- The second quarry is derived from Chatt, the exploitation is carried out by explosive. The waste from this quarry is (marble of different granular classes), downgraded powder (**Figure 2**), the waste rate is 19% of production.

The valorization of this waste in the manufacture of mortars and concretes remains in the current state of investigation. Several studies have been interested by the feasibility of partially replacing an ordinary aggregate with a marble aggregate,

Figure 1.
Waste from the block quarry.

Figure 2.
Downgraded powder from Chatt.

or partially replacing cement with marble fillers. Hebhoub and Belachia [1] studied the valorization of marble waste aggregates in the concrete composition with total and partial incorporation from 0 to 100%, they found an improvement of compressive and tensile strength as well the workability of the concrete. Binici et al. [2] examined the incorporation of marble coarse aggregates in concrete and they found that it tends to decrease its chloride penetration, reaching a 70% reduction compared to standard concrete at 28 days of immersion. Hasan and Ahan [3] found that, replacing standard sand by the marble dust, with percentages of 15–75%, leads to an increase of compressive and tensile strength from 20 to 26% and 10 to 15%, respectively. Aliabdo et al. [4] evaluated the possibility of reusing marble dust as a partial replacement of cement and sand in concrete; the results found indicate an improvement in the physical and mechanical properties of concrete. Djebien et al. [5] reused marble waste as sand in self-compacting concrete, they found that substitution of marble waste reduces density and air content, and ensures cohesion and resistance to segregation.

Gesoğlu et al. [6] concluded that the incorporation of marble dust in concrete decreased chloride penetration. In particular, replacing 5% of marble dust, by cement weight, led to the highest decrease in chloride penetration. Corinaldesi et al. [7], studied also the effect of marble powder in concrete, concluding that, up to given ratios of replacement, concrete durability can be improved. Belaidi et al. [8] examined the effect of the substitution of marble powder on the properties of self-compacting concrete, in percentage different from 10 to 40% they have been shown an improvement on the workability of concrete with a negative effect on compressive strength. Aruntas et al. [9] studied the addition of marble dust in cement production, obtaining very similar results to the ones observed in our study. Chavhan and Bhole [10] produced concrete mixtures by replacing gravel with marble powder, the rate varied between 5 and 50% they found enhancement in compressive and tensile strength of samples with 50% of marble.

2. Used materials

2.1 Cement

Cement CEM I class 42.5 of the origin of the Ain kbira-Sétif cement plant (East of Algeria) with an absolute density of 3.22 g/cm^3 and a Blaine specific surface of 3000 cm^2/g. Physical properties; chemical and mineralogical composition of cement used in this work is presented in **Table 1**.

2.2 Sand

Three types of sand were used in this work, the first is sea sand, it is a nature rolled, class 0/2 of origin Larbi ben Mhidi-Skikda and the second one is a dune sand of nature rolled class 0/1 of origin Wadi Zhor-Skikda. We are used a mixture between the both with similar quantities (natural sand). The third sand is a marble waste sand from the Filfila quarry—Skikda class 0/2 (discarded powder exposed to the weather). The physical and chemical properties of different sands used are presented in **Table 2** and the particle size curves are given in **Figure 3**.

2.3 Water

Potable water was used in all the mixes and curing of the specimens. (temperature was between 20 ± 2°C).

Designation	Results (%)
Density (g/cm^3)	3.07
Specific surface Blaine (cm^2/g)	3700
CaO	65.85
Al_2O_3	4.13
Fe_2O_3	4.16
SiO_2	21.31
MgO	1.34
Na_2O	0.16
K_2O	0.25
Cl	0.003
SO_3	2.13
Free CaO	0.5
MS	2.43
MAF	0.88
C3S	72.25
C2S	8.83
C3A	3.14
C4Af	14.7

Table 1.
Physical properties and chemical and mineralogical composition of the cement.

Designation	Sea sand	Dune sand	Marble waste sand
Physical properties			
Apparent density (g/cm^3)	1.606	1.850	1.50
Absolute density (g/cm^3)	2.570	2.60	2.700
Value of blue methylene (%)	0.25	0.7	0.35
Sand equivalent (%)	81	84	67.11
Absorption	1.10	1.15	2.30
Fineness modulus (%)	2.83	1.88	1.65
Fines content (%)	1	2.5	6
Chemical composition			
$CaCO_3$	—	—	98.67
CaO	4.01	0.80	55.29
Al_2O_3	0.76	2.36	0.14
Fe_2O_3	1.17	1.15	0.09
SiO_2	87.32	94.09	0.53
MgO	0.15	0.14	0.2
Na_2O	0.090	0.20	0.00
K_2O	0.200	0.58	0.01
Cl^-	0.00	0.00	0.025

Designation	Sea sand	Dune sand	Marble waste sand
SO_3	0.01	0.01	0.04
PF	—	—	43.40
Insoluble residue	—	—	0.035

Table 2.
Physical and chemical properties of different sands.

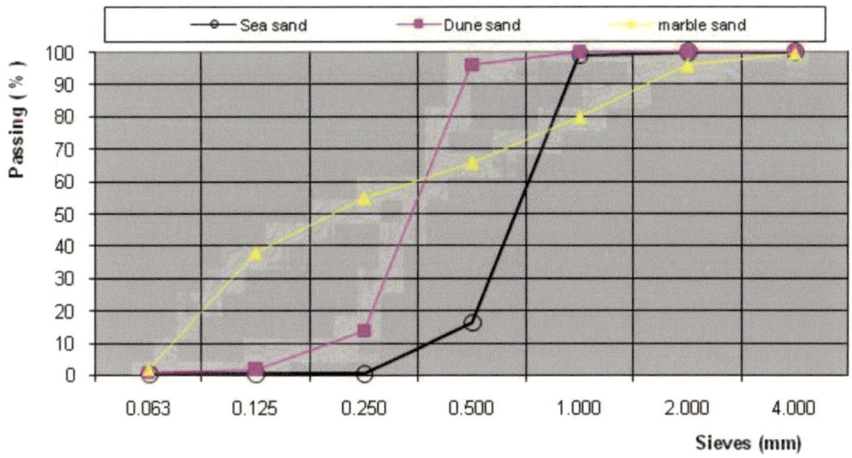

Figure 3.
Granulometric curves of the various sands.

From the results obtained we can draw the following observations:

- The absolute density of marble waste sand is higher than that of natural sand.
- Marble waste sand is less clean than natural sand.
- The marble waste sand has the highest percentage of fine clayey.
- The absorption coefficient of marble waste sand is stronger than that of natural sand.
- Marble waste sand is rich in calcium carbonate while the essential component of natural sand is silica.

3. Experimental program

The objective of this work is to study the effect of partial substitution of natural sand (50% dune sand and 50% sea sand) by a marble waste sand with rates 5, 10, 15 and 20% on the characteristics of mortars.

Reference mortar (control mortar) was prepared (0% marble waste sand) according to standard EN 196–1, with a quantity of water adjusted in order to obtain a reference consistency, the fixed parameters are cement and water dosage. The different mixes were prepared by replacing four percentages of natural sand with the same mass percentages of marble waste sand. This makes a total of five different mixes, including a control mortar. $4 \times 4 \times 16$ cm^3 prismatic test pieces were made to determine the performance of hardened mortars as well as $5 \times 5 \times 5$ cm^3 test

Figure 4.
Prismatic test pieces (4 × 4 × 16).

Figure 5.
Cubic test pieces (5 × 5 × 5).

pieces for acid attack tests **Figures 4** and 5. Specimens produced from fresh mortar were demolded after 24 h and were then cured in water at 20 ± 2°C until the date of the test. All tests are realized in the same conditions (laboratory conditions).

The different compositions of the mixtures for the five formulations are given in **Table 3**.

Notation	Dune sand (g)	Sea sand (g)	Marble waste sand (g)	Cement (g)	Water (ml)
CM (0%)	675	675	0	450	252
M (5%)	641.25	641.25	67.5	450	252
M (10%)	607.5	607.5	135	450	252
M (15%)	573.75	573.75	202.5	450	252
M (20%)	540	540	270	450	252

Table 3.
Compositions of mixtures.

4. Tests performed

The tests carried out on the different formulations are:

- Consistency, measured by the mini slump test in accordance with standard NF EN 1015-3.
- Density in the fresh state according to standard NF EN 1015-6.
- Air content determined according to standard NF P 18-353.
- Flexural tensile and compressive strength at the age of 2, 7, 28 and 90 days, measured on $4 \times 4 \times 16$ cm^3 prismatic specimens preserved in water in accordance with standard EN196-1.
- Shrinkage and mass loss on specimens $4 \times 4 \times 16$ cm^3 according to NF P 18-433.
- Absorption by immersion measured by Neville, 2000.
- Acid attack measured on cubic specimens of size $5 \times 5 \times 5$ cm^3 according to ASTM C-267-96 standard.

5. Results and discussion

5.1 Consistency of mortars

The slump of various mortars is evaluated according to the standard NF EN 1015-3. The test is mentioned in the **Figure 6**.

The results of consistency of different mortars are shown in **Figure 7**.

The figure shows that the increase in marble waste sand rate in the mortars increases slightly the consistency of the mortar. The maximum value is recorded by the mortar with 15%, with an increase of around 20% compared to that of the control mortar. This trend can be explained by the amount of fines present in the marble waste sand which enter the pores and thus release the trapped water, which results in better consistency.

Figure 6.
Consistency of mortar.

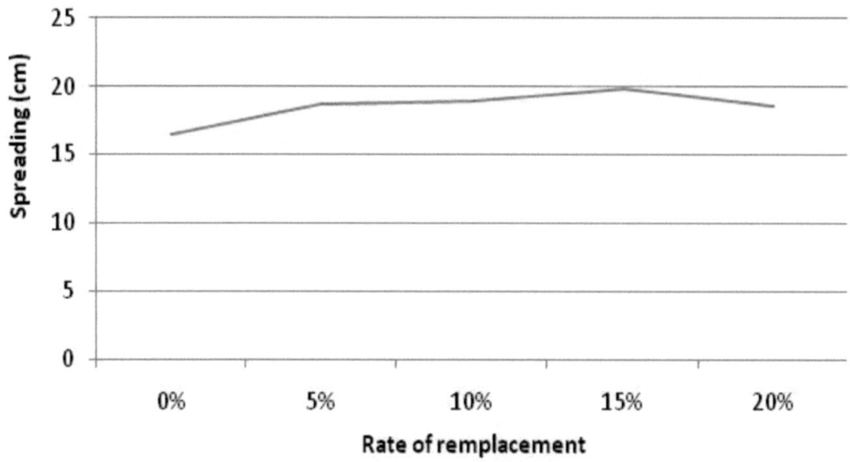

Figure 7.
Variation of consistency versus substitution rate.

5.2 Density of mortars

The density of various mortars is evaluated according to the standard NF EN 1015-6. The results of the density of mortars are shown in **Figure 8**.

The results show an increase in the density of all mortars. This growth is clearer with the increase in the rate of marble waste sand. The density values increase from 2.168 kg/m^3 for the reference mortar to 2.175, 2.192, 2.201, and 2.186 kg/m^3 for the mortars containing 5, 10, 15 and 20%. The increase in the density of the mortar to 15% of marble waste sand is around 1.52%. It is mainly due to the higher density of marble waste sand, which is higher than that of natural sand (dune sand and sea sand), and also to the retention of water by the grains of marble waste during the mixing.

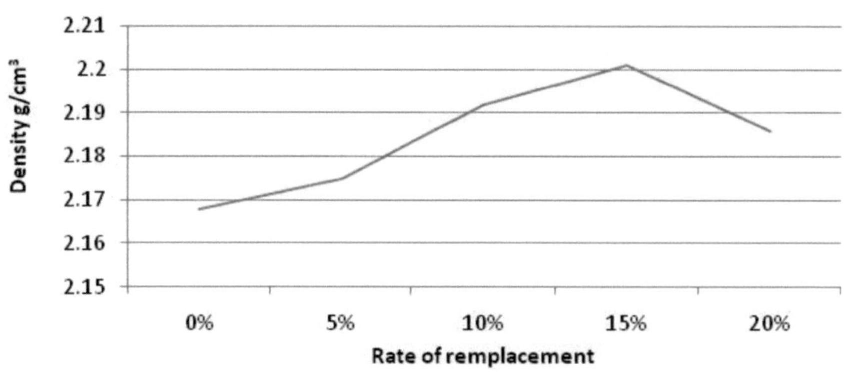

Figure 8.
Variation of density versus substitution rate.

5.3 Air content

The introduction of marble waste sand (**Figure 9**) leads to a decrease in the air content regardless of the substitution rate. The volume of occluded air decreases slightly from 7.4% for the reference mortar to 4.8% for the mortar incorporating 15% of marble waste sand. The significant reduction in the volume of air entrained in composite mortars of marble waste sand is related to the increase in the

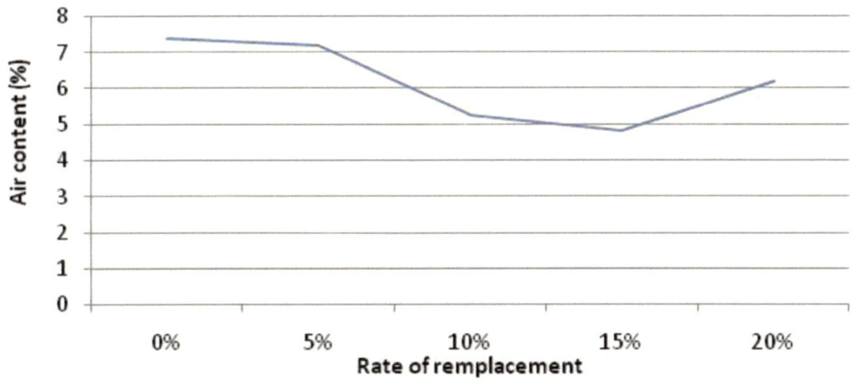

Figure 9.
Change in air content as a function of the substitution rate.

compactness of the mixtures following the substitution of natural sand by marble waste sand [11].

5.4 Compressive and flexural tensile strength of mortars

After 2, 7, 28 and 90 days of water curing, the 4 × 4 × 16 cm^3 samples were used for compressive and flexural tensile strength tests. The results are shown in **Figures 10** and **11**, respectively.

At early ages (2 and 7 days), and through **Figure 10**, the compressive strengths of mortars based on marble waste sand presented higher strength than that of control mortar, except for mortar with 5%. The best gain of compressive strength is of the order of 13.65 and 35.2%, it is recorded in the mixture of (20%) on the two deadlines respectively. These improvements in both levels can be explained by the presence of calcium carbonate [12, 13], which favor the creation of nucleation sites and hence the formation of calcium carbo-aluminates.

The compressive strength of specimens with marble waste sand at rates above 5% are better than that of the control mortar and the best gain is noted in the mixture of (20%) substitution rates. In fact, above 5% of the substitution rate of marble waste sand, all mortar compressive strengths are increasing, due to the increased compactness of mortars with less occluded air and constant W/C ratio [14].

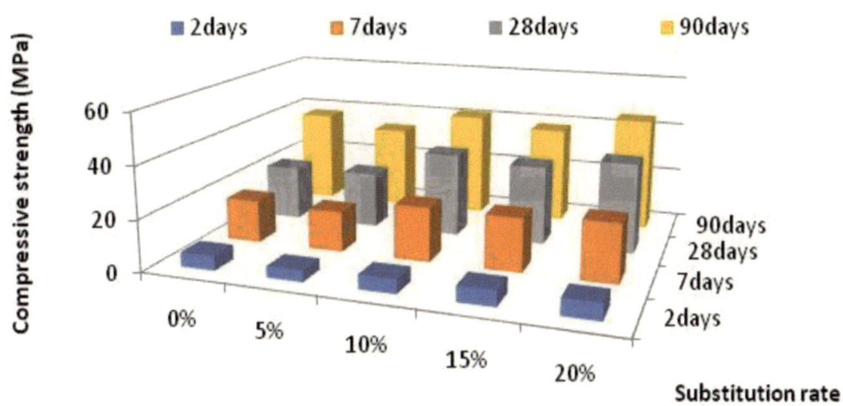

Figure 10.
Effect of the substitution rate on the compressive strength.

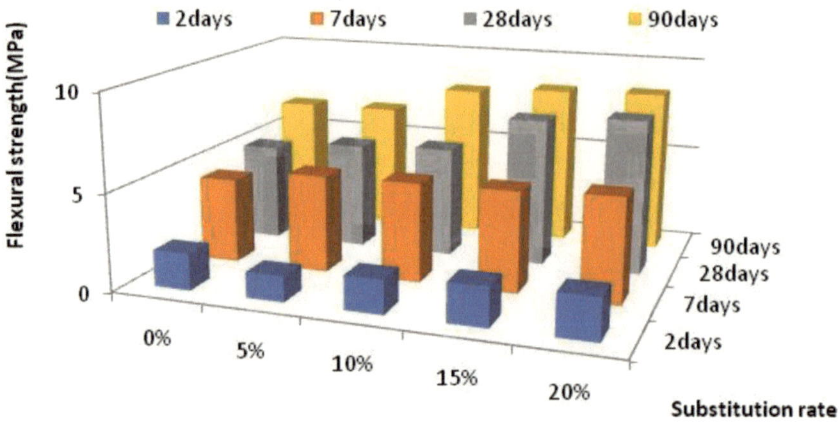

Figure 11.
Effect of the substitution rate on the flexural tensile strength.

Generally, we note that the flexural tensile strengths of the mortars with marble waste sand, at all the ages (7, 28 and 90 days) are better than that of control mortar. The most significant value is achieved with the 20% mixture of marble waste sand, which presents a gains of 13.29, 22, 58 and 28% compared to the control mortar on the ages 2, 7, 28 and 90 days respectively. Two factors can explain these notations. Mortars based on marble waste sand contain quantities of fine particles, which favor granular stacking during mixing and thus causes an increase in flexural strength. However, the marble waste sand is characterized by more acute and porous grains, so that the bond with the cement paste of the mixture is better [7].

5.5 Absorption by immersion

The absorption of water by immersion is a property related to the durability of mortar, it allows estimating the volume of open pores of specimens by the penetration of water through the structure of these pores.

When the ratio of replacement of marble waste sand in mortar increased (**Figure 12**), there was an increase in water absorption, especially for the 20% of

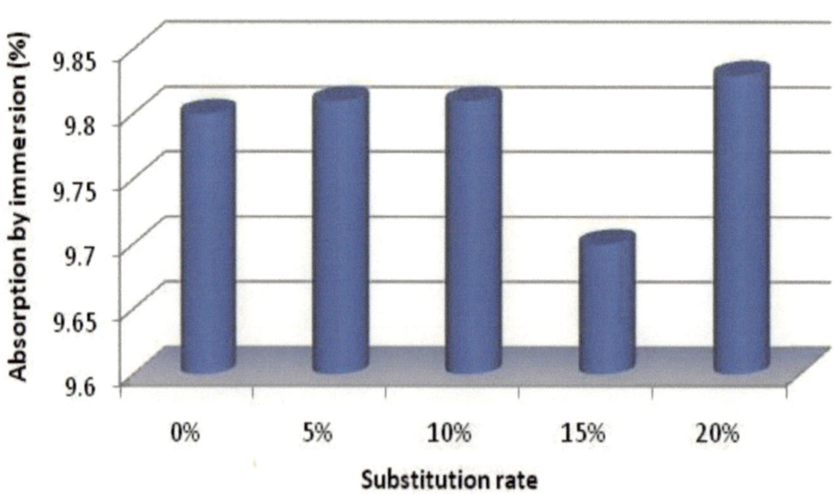

Figure 12.
Absorption of water by immersion as a function of substitution rate.

replacement, due to the height porosity of specimens with marble waste sand. Also, it can be assumed that that the matrix-sand interface of marble waste often gathers pores thus increasing porosity.

5.6 Weight loss

The results of the mass loss of the various mortars are presented in **Figure 13**.

From the **Figure 13**, it is clear that the weight loss of mortars is influenced by the incorporation of marble waste sand. The values obtained are generally high compared to the control mortar. The optimum value is found in the 15% marble waste sand mortar. This is due to the departure of the water initially retained by the grains of the marble sand.

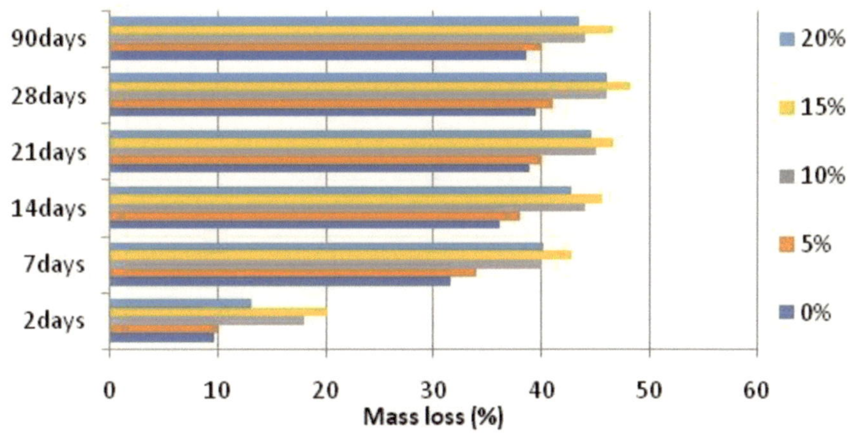

Figure 13.
Weight loss of different mortars.

5.7 Shrinkage

The shrinkage test is carried out according to standard NF P 18-433, test is mentioned in **Figure 14**.

Figure 14.
Shrinkage measurement on prismatic test piece.

Shrinkage results of various mortars are presented in **Figure 15**.

Figure 15 shows that the incorporation of marble waste sand has a considerable impact on the shrinkage of mortars stored in the laboratory room with a relative humidity of 80%. The mean values recorded are higher than that displayed for the control mortar, except for the mortar of 5% marble waste sand from 2 days until 21 days, where the mortar has a similar shrinkage to the mortar witness.

The highest value is found in the mortar with 20% marble waste sand on all ages. These expected shrinkage values are due to the evaporation of the free water contained in the test specimens.

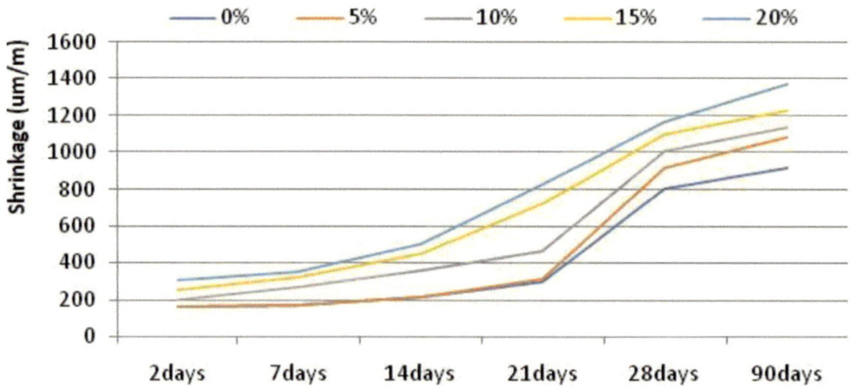

Figure 15.
Shrinkage results of various mortars.

5.8 Acid attack

After 28 days of water curing, the $5 \times 5 \times 5$ cm^3 specimens were immersed in two solutions, HCl and H_2SO_4 acid with the same concentration 5% (**Figure 16**). The aggressive solutions were renewed every 14 days. After 1, 7, 14, 21, 28, 56 and 90 days, they were used to estimate the weight loss according to the standard ASTMC267-96.

Results of the weight changes for the different mortars preserved in HCl and H_2SO_4 solution are presented in **Figures 17** and **18**.

The curves, presented by **Figure 17**, show the weight loss in % measured at the end of each aging of mortars stored in HCl solution (after 28 days of cure in water).

Figure 16.
Specimens immersed in acid solution.

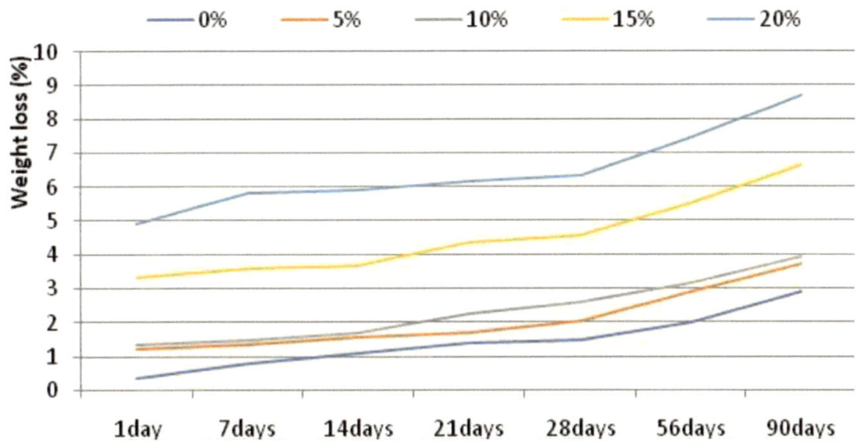

Figure 17.
Weight loss of specimens after 1–7–14–21–28–56–90 days of immersion in 5% HCl.

It can be observed that all specimens had a weight loss, through the reaction between the calcium hydroxide Ca(OH)2 and the chlorine, the reaction was expressed in Eq. (1).

$$Ca(OH)2 + 2HC \rightarrow CaCl_2 + 2H_2O \tag{1}$$

Calcium chloride ($CaCl_2$) causes dissolution of cement, the production of $CaCl_2$ salt increases the porosity because it is very soluble in water.

It can be also noted that in all ages the loss weight increased proportionally with marble waste sand substitution, and the control mortar presented the lowest weight loss at all ages. It can be concluded that a high calcium carbonate ($CaCO_3$) content in marble waste sand increases the capacity of mortars to react with aggressions according the reaction Eq. (2)

$$CaCO_3 + 2HCl \rightarrow CaCl_2 + H_2O + CO_2 \tag{2}$$

A sudden loss of weight was noticed during 1–90 days (**Figure 18**). It can be also noted that in early age all weight loss is close, the highest value of weight loss

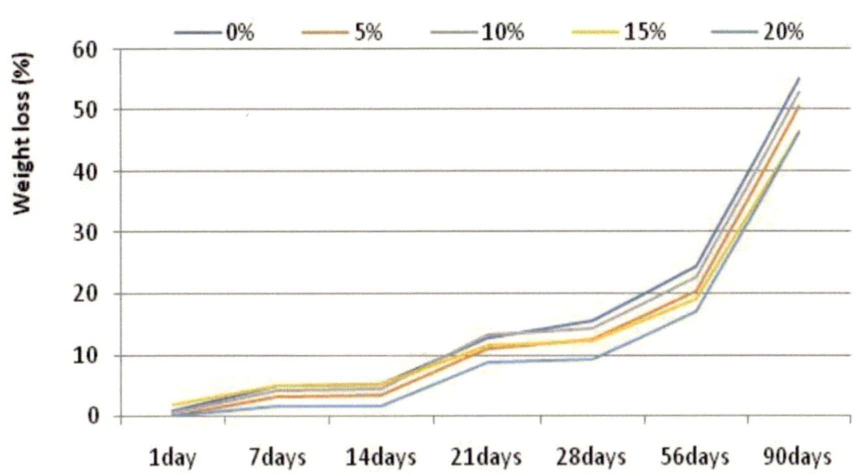

Figure 18.
Weight loss of specimens after 1–7–14–21–28–56–90 days of immersion in 5% H_2SO_4.

corresponds to control mortar, when the mortars are attacked by sulfuric acid H_2SO_4, they react with the Portlandite Ca(OH)2 resulting from the hydration of the cement [15], which causes the of gypsum. The process is described by the following chemical reaction:

$$Ca(OH)2 + H_2SO_4 \rightarrow CaSO_4.2H_2O \quad (3)$$

The low percentage of alumina Al_2O_3 in marble waste sand decreases the formation of C3A, which reacts with gypsum to produce ettringite (Eq. 4), that is why the increase in marble waste sand percentage decreases the weight loss of mortars.

$$3CaSO_4 + 3CaO.Al_2O_3.6H_2O + 26H_2O \rightarrow 3CaO.\ Al_2O_3.\ 3CaSO_4.32H_2O \quad (4)$$

6. Conclusions

After the comparative study between mortars based on natural sand and marble waste sand, we can draw the following conclusions:

- The optimal density is observed for the mortar based on 15% of marble waste sand, due to the actual density and the water retention by the grains of marble waste.

- The best consistency is given by the mortars based on 15% of marble waste sand, the presence of fines promotes the slump of the mortars.

- The mortar with 15% marble waste sand registers the low volume of air content, which is explained by a better compactness.

- Compressive and flexural strength, in all ages, of mortars based on marble waste sand are better than those of mortars based on natural sand. The mortar with 20% of marble waste sand is the most effective.

- The absorption of water of mortars containing marble waste sand is high compared to that of the control mortar. This is caused by the large volume proportion of the capillary pores.

- The mortars with the marble waste sand approved a greater shrinkage than that of the control mortar, following the evaporation of the free water existing in the test specimens.

- Concerning acid attack, and in HCl solution, the mortar with marble waste sand had a bad resistance to aggression due the high amount of $CaCO_3$ in marble. On the other hand, marble sand enhances resistance to H_2SO_4.

In light of the analysis, the conclusion of this study is that the introduction of marble waste sand was beneficial to some properties and durability of mortars.

Author details

Hebhoub Houria[1*], Kherraf Leila[1], Abdelouahed Assia[1] and Belachia Mouloud[2]

1 Department of Civil Engineering, LMGHU Laboratory, University of Skikda, Algeria

2 Department of Civil Engineering and Hydraulic, LMGHU Laboratory, University of Guelma, Algeria

*Address all correspondence to: hebhoubhouria@yahoo.fr

References

[1] Hebhoub H, Belachia M. Use of the marble wastes in the hydraulic concrete. Nature and Technology. 2011;**04**:41-46

[2] Binici H, Shah T, Aksogan O, Kaplan H. Durability of concrete made with granite and marble as recycled aggregates. Journal of Materials Processing Technology. 2008;**208**(1): 299-308

[3] Hasan S, Ahan A. Recyclability of waste marble in concrete production. Journal of Cleaner Production. 2016;**131**: 179-188

[4] Aliabdo AA, Abd Elmoaty AEM, Auda EM. Re-use of waste marble dust in the production of cement and concrete. Construction and Building Materials. 2014;**50**:28-41

[5] Djebien R, Hebhoub H, Belachia M, Berdoudi S, Kherraf L. Incorporation of marble waste as sand in formulation of self-compacting concrete. Structural Engineering and Mechanics. 2018;**67**(1): 87-91

[6] Gesoğlu M, Güneyisi E, Kocabag ME, Bayram V, Mermerdas K. Fresh and hardened characteristics of self compacting concretes made with combined use of marble powder, limestone filler, and fly ash. Construction and Building Materials. 2012;**37**:160-170

[7] Corinaldesi V, Moriconi G, Naik T. Characterization of marble powder for its use in mortar and concrete. Construction and Building Materials. 2010;**24**(1):113-117

[8] Belaidi ASE, Azzouz L, Kadri E, Kenai S. Effect of natural pozzolana and marble powder on the properties of self-compacting concrete. Construction and Building Materials. 2012;**31**:251-257

[9] Aruntas HY, Gürü M, Dayı M, Tekin I. Utilization of waste marble dust as an additive in cement production. Materials and Design. 2010;**31**: 4039-4042

[10] Chavhan PJ, Bhole SD. To study the behaviour of marble powder as supplementary cementitious material in concrete. International Journal of Engineering Research and Applications. 2014;**4**(1):377-381

[11] Djebien R, Belachia M, Hebhoub H. Effect of marble waste fines on rheological and hardened properties of sand concrete. Structural Engineering and Mechanics. 2015;**53**(6):1241-1251

[12] Bachiorrini A, Cussino L. Hydratation du ciment alumineux en présence d'agrégats calcaires. In: 8th Proceedings of the International Congress, Chemistry of Cement, RIO DE JANEIRO, IV; 1986. pp. 383-388

[13] Hebhoub H, Belachia M, Djebien R. Introduction of sand marble wastes in the composition of mortar. Structural Engineering and Mechanics. 2014; **49**(4):491-498. DOI: 10.12989/ sem.2014.49.4.491

[14] De Larrard F. Construire en béton, l'essentiel sur les matériaux. Paris: Presse de l'école nationale des Ponts et chaussées; 2002. ISBN 2-85978-366-0. p. 191

[15] Abdelouahed A, Belachia M, Sebbagh T. Effect of SCMs on mechanical, chemical and microstructural properties of SRC in acidic medium. European Journal of Environmental and Civil Engineering. 2016;**22**(2):212-225. DOI: 10.1080/ 19648189.2016.1185971

Chapter 3

Use of Waste Foundry Sand (WFS) as Filler in Hot-Mixed Asphalt Concrete

Nilton de Souza Campelo, Karine Jussara Sá da Costa, Raimundo Kennedy Vieira and Adalena Kennedy Vieira

Abstract

The environmental issue has become a topic of relevant discussion in modern society, given the current awareness that construction inputs are finite, and a large amount of waste can be reused as a building material in engineering works. The products used in foundry industry can be non ferrous and ferrous and the residue produced by this last one is not potentially hazardous to human health. The waste foundry sand (WFS) fits this reuse and can be employed in asphalt mixtures, in partial or complete replacement of the conventional filler, i.e. Portland cement (PC). In this sense, this work analyses five asphalt mixtures, one using 100% CP (reference mixture) as filler, and the other four using WFS in proportions of 25-100%, every 25% of the total amount, in 5% (in mass) of the maximum replacement. The mixtures were physically and mechanically characterised according to the Marshall methodology and subsequently submitted to the tests of static indirect tensile strength (static ITS), resilient modulus (RM), repeated-load indirect fatigue (fatigue life) and unconfined static creep. The results of the tests showed that all mixtures with WFS residue presented physical and mechanical parameters within Brazilian standards following the Marshall methodology.

Keywords: waste foundry sand (WFS), mineral filler, asphalt mixture, fatigue life, static creep, resilient modulus, static indirect tensile strength, Marshall stability

1. Introduction

Industries annually generate millions of metric tons of solid by-products, and most of these materials have been landfilled at considerable cost since. Modern society has been developing beneficial reuse of industrial by-products in a variety of applications [1–3]. Recycling of waste construction materials saves natural resources, saves energy, reduces solid waste, reduces air and water pollutants and reduces greenhouse gases [4, 5]. The transportation, construction and environmental industries have the greatest potential for reuse because they use vast quantities of earthen materials annually. Replacement of natural soils, aggregates and cements with solid industrial by-products is highly desirable [1, 2].

The steel industry produces a myriad of metal components for industrial chains such as the automobile industry, which in turn generates mineral discarded sand moulds (waste foundry sand/WFS) that end up occupying large volumes in

IntechOpen

landfills [6]. The major portion of the WFS is considered as non-hazardous waste and is currently deposited in a special WFS landfill that is remote from areas of settlement [7–10].

The metal casting industry annually discards about 10% of foundry sand for production, i.e. approximately an estimated 9–10 million tons of WFS each year, in the USA [5, 10, 11]. Generally speaking, approximately 1 ton of foundry sand is needed to produce 1 ton of metal casting [8, 12]. WFS can be used as an alternative material (fine aggregate in asphalt mixtures) in highway constructions allowing the increasing of the lifespan of landfills [13].

This work analyses the physical and mechanical behaviour of asphalt mixtures, using the WFS as a mineral filler in asphalt concrete, in 5% (in mass) of maximum replacement to conventional Portland cement (CP). The waste was obtained from an industry located in the free-trade zone of Manaus city, Amazon State, Brazil. The results showed that the addition of industrial WFS in asphalt mixture resulted in adequate performance of the mixtures.

2. Literature review

Waste foundry sand is generated by industries that use sands, binders and additives to form moulds and cores for castings. Sands are chosen for several reasons; they are readily available everywhere, inexpensive, highly refractory and readily bonded by clays or other inorganic and organic materials [8, 9, 14]. The mould forms the outside of the castings; the core forms the internal shape. When the part to be made has deep recesses or hollow portions, sand cores must be provided in the mould [3]. The material to be used to form moulds and cores in a foundry should have cohesiveness and porosity properties at the same time. Adding binder (bentonite, resins, cement, sodium silicate and oils) will improve the cohesiveness of the sand grains but will tend to reduce porosity. Additives are those materials which are added to the bonded sands to improve properties, either during the moulding process or during the casting process or both [8]. The moulding processes which involve sand are (1) green sand moulding (or clay-bonded sand, [12]), (2) chemically bonded process and (3) shell moulding process [3, 5, 8, 9]. The most commonly used process is green sand moulding [15]. Green sand is composed of four major materials. Sand comprises 85–95% of the green sand mixture. Most often the sand is inert silica, but olivine and zircon sand are also used [8, 15–17]. Approximately, 4–10% of the mixture is made of some form of clay, e.g. bentonite. The clay acts as a binder for the green sand and provides strength and plasticity. Combustible additives like sea coal, cereal, fuel oil and wood flour typically make up from 2 to 10% of the green sand mixture. The final additive of green sand is water which is usually added in small percentages (2–5% by weight) [5, 8]. Chemically bonded sands are those that use furan, phenolic urethane and acid cured no-bake systems, as well as alkyd and phenolic urethane cold box processes. Shell moulding uses a mixture of sand and thermosetting resin (usually phenol formaldehyde) to form the mould [8, 17].

The physical, chemical and mechanical characteristics of virgin sand make it a popular material for construction engineering, but after several reuses in moulds and cores, it becomes WFS [7]. The grain size distribution of WFS is quite uniform, with a majority of the sizes (85–95%) falling within a narrow range between 0.6 and 0.15 mm, and 5–12% is smaller than 0.075 mm [5, 8, 10] or between 1 and 16.5% [14]. According to Tikalsky et al. [17], more than 80% of the particles by mass are concentrated by size between 0.15 and 0.70 mm, compared to 0.30–4.75 mm for conventional fine aggregate. Most of the WFS materials reported are found to be medium to fine sand. WFS have been found to be too fine to satisfy the specifications for general fine aggregate [8, 10, 12]. WFS has uniform equidimensional subangular to rounded grains, and a few has rounded grains [8, 10, 17, 18].

For density and unit weight, the values found for the WFS were very close to conventional aggregate [13]. The bulk specific gravities reported in the literature on WFS ranged from 1.985 to 2.722 [8, 17]. In most of the cases WFS have been reported to be almost dry. The moisture content as received for WFS were reported to be in the range of 0.0–4.85% [8, 17, 18]. Concerning absorption, the values are relatively higher than those obtained for the natural aggregate, due to the presence of organic matter [6, 8]. The percentage absorption values on WFS samples have been reported to vary between 0.3 and 6.2% [3, 8, 17].

Over the past three decades, there have been several studies around the world on the use of WFS in engineering works, in different areas: base and subbase layers of highway construction [19–21], embankments [22, 23], hydraulic barriers [24], asphalt mixtures [3, 7, 16, 25, 26], etc.

Highway subbase layers using WFS have been shown to resist winter conditions (freeze–thaw cycles) better than specimens of reference materials [5, 17, 19]. If a subbase layer stabilised with WFS is compacted in field at dry of optimum content then it will have an increase in its strength [19, 20, 27].

It has been mentioned in the literature that the fines of WFS affect the properties of asphalt concrete negatively [7, 28]. The amount of WFS used in an asphalt mixture depends largely on the amount of fines in the WFS [5, 12, 14, 29]. Studies have recommended that WFS should replace successfully as much as 15% (in mass) of the conventional sand (fine) content in asphalt concrete [3, 9]; 8–10, 10–20 and 10%, respectively, in engineering practice in Pennsylvania, Michigan and Tennessee States [5]; 35% [30]; 15% [26]; 10% [7, 13]; 15% [10, 31]; 35% [27, 32]; and 15–30% [14].

Concerning physical characteristics, the densities of the mixtures decreased as the percentage of WFS in the asphalt concrete increased [7, 9, 10, 12, 13, 17, 32]. Percentage of air voids and voids in the mineral aggregate (VMA) were found to increase with blending of increased quantities of WFS [8, 9]. The optimum asphalt content (4.9–6.8%) for HMA mixtures containing various amounts of foundry sand is comparable to the content of mixes not containing foundry sand [14, 17]. The OAC increases with increase in the WFS percentage [13], although Miller et al. [14] found lower values for mixtures containing WFS, in relation to control ones. According to this author, the mixtures obtain the higher percentage of OAC with the WFS with the higher amount of particles passing the #200 sieve. This happens due to the fineness properties of material and increase of surface area [10, 32].

Regarding the mechanical characteristics, the Marshall stability of the asphalt concrete samples containing WFS decreases as the quantity of WFS is increased [3, 6–8, 10, 12, 29, 32]. The flow values of mixtures decreased with increasing percentage of WFS in the asphalt concrete mixtures [7,–10, 13]. The indirect tensile strengths of the asphalt cement mixtures decreased as the percentage of WFS material was increased [7–10, 12, 13, 32]. However, Abdulsattar and Mohammed [25] found that all the WFS mixtures that they analysed showed higher tensile strength than the control mixture. According to Tikalsky et al. [17], the level of air voids and saturation greatly influenced the indirect tension values.

In relation to moisture susceptibility, WFS has little effect on top-down fatigue cracking resistance and moisture susceptibility of the mixtures [32]. When WFS replacement is higher than 15%, asphalt mix may become more sensitive to moisture damage (i.e. stripping) due to the presence of silica [10, 27]. WFS, on average, decreases the unconditioned tensile strength and thus the durability of asphalt mixtures; on the other hand, WFS do not necessarily increase or decrease a mixture's rutting potential but do improve fatigue performance [17].

3. Materials and methods

3.1 Origin of materials

The experimental procedure of this research contemplates the dosage and physical and mechanical tests on five hot-mixed asphalt concrete (HMAC) mixtures using the conventional Portland cement filler (as reference) and four other mixtures using WFS, replacing the cement gradually in proportions of 25%. This residue was produced by the foundry industrial process of a company located in free-trade zone of Manaus city, Amazon State, Brazil, which produces clutch assembly lines (pressure and friction plates, discs, outer housing, etc.) for the motorcycle industry. **Figure 1a** shows one of the several kinds of pieces that are produced in that industry, while **Figure 1b** presents the WFS studied. The annual production of WFS in that industry was about 1500 tons in 2014 (SUFRAMA, 2016). The coarse aggregate (natural pebble) came from the "Japurá" River (an Amazon River affluent) riverbeds and was extracted by dredging, but it was acquired in the local market. The fine aggregate (clean sand) came from mining extraction in the vicinity of the city (about 30–50 km), but it was acquired in the local market as well. The mineral filler used was Portland cement II-Z-32 type. Finally, asphalt cement (AC) 50/70 grading was used, produced by the oil refinery of Manaus (REMAN). The materials used in this research and their respective origins are listed in **Table 1**.

3.2 Characterisation of materials

All mineral aggregates used in the asphalt mixtures were tested according to the standards described in **Table 2**, mainly by the Brazilian highway standards, which are most similar to known international standards. In relation to the asphalt cement (AC—50/70 penetrating grading), it was submitted to complete characterisation according to standards shown in **Table 3**.

In order to avoid the presence of impurities, the residue was washed in sieves Nos. 200, 300 and 400, before subjected to characterisation tests and used in asphalt mixtures. The WFS filler was subjected to chemical analysis (XRF) made by an X-Ray spectrometer equipment (720 energy dispersive, Shimadzu), through drying and subsequently pressing the sample in a disc form. The equipment can perform analyses from sodium to uranium, has a rhodium tube and is cooling by liquid nitrogen. Besides that, the WFS filler was also submitted to the X-Ray diffraction (XRD) in order to be characterised its crystalline phases. The equipment used in the analysis was

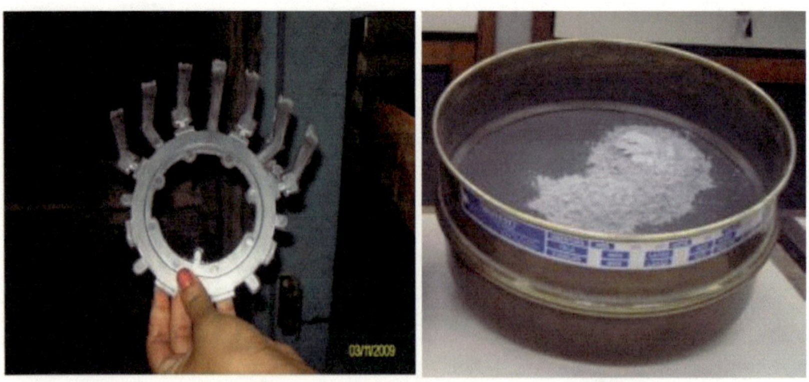

a) b)

Figure 1.
(a) A piece (to be deburred) produced at the trade zone of Manaus city industry. (b) WFS to be tested.

Material	Origin
Sand	Market of Manaus
Pebble	Market of Manaus
Portland cement (PC) II-Z-32 (mineral filler)	Market of Manaus
Asphalt cement (AC) (50/70 grading)	Oil refinery of Manaus (REMAN)
Waste foundry sand (WFS)	Industry of free-trade zone of Manaus

Table 1.
Provenance of HMAC component materials.

Material	Brazilian standard	Title	Acceptance parameters (Brazilian standard)	Similar international standard
Pebble	NBR NM 53/2009	Coarse aggregate—determination of the bulk specific gravity, apparent specific gravity and water absorption	Greater than 0.88 and 2.00 g/cm^3; less than 18%, respectively	ASTM-T-85
Pebble	NBR NM 51/2001	Coarse aggregate—test method for resistance to degradation by Los Angeles machine	Less than 50%	AASHTO-T-96
Pebble, Sand, Fillers	NBR NM 248/2003	Aggregates—sieve analysis of fine and coarse aggregates	Within granulometric range	ASTM-C136/ C136M-14
Pebble	NBR 12583/1992	Coarse aggregate—coating to bituminous binder	Qualitative test (visual analysis)	—
Sand	NBR NM 52/2009	Fine aggregate—determination of the bulk specific gravity and apparent specific gravity	Greater than 1.60 and 2.60 g/cm^3, respectively	ASTM-C128–01
Fillers	NBR NM 23/2001	Portland cement and other powdered materials—determination of density	Greater than 3.00 g/cm^3	ASTM-C188–09
WFS	NBR 16137/2010	Non-destructive testing—material identification by spot test, X-ray fluorescence spectrometry and optical emission spectrometry	—	ASTM-C114–15
WFS	—	Wavelength dispersive X-ray fluorescence spectrometry	—	ASTM-C1365

Table 2.
Aggregate characterisation tests.

the D8 Focus-Bruker diffractometer, with monochromatic cuprum radiation (CuKα, λ = 1.5418 Å), operating at 35 kV and 40 mA. A laser particle size analyser was used to determine with precision the particle size of both mineral fillers (PC and WFS).

3.3 Dosage method of the SMA mixtures

Since the tests were performed 10 years ago, asphalt concrete studies were developed through the traditional Marshall method and not by current Superior

Brazilian standard	Test	Unity	Similar international standard	AC 50/75
NBR 14756	Apparent specific gravity, 25°C	g/cm^3	AASHTO T 228	1.010
NBR 6576	Penetration, 25°C, 100 g, 5 s	0.1 mm	AASHTO T 49	58
NBR 14950	SSF viscosity, 135°C	s	AASHTO T 72	160
NBR 15184	Brookfield viscosity, 135°C, sp21, rpm 20	cP	AASHTO T 316	286
NBR 6560	Softening point	°C	AASHTO T 53	53
NBR 6293	Ductility, 25°C	cm	AASHTO T 51	>120

Table 3.
Properties of asphalt cement (AC—50/70 penetrating grading) used in the mixtures.

Performing Asphalt Pavements (Superpave) methodology. After the characterisation of all components of asphalt concrete, the materials were classified in the "C" granulometric range limits of Brazilian highway specifications following the Marshall dosage method, as shown in **Figure 2**. The curves obtained fitted in the area defined by the two curve limits of the "C" range, minimum and maximum. After fixing the particle size distribution of aggregates of the mixture, the probable optimum asphalt content (OAC) was estimated by the expression derived from the work of Duriez (1950) based on the specific surface of the aggregates:

$$S = \frac{0.17G + 0.33g + 2.30A + 12a + 135f}{100} \tag{1}$$

where S is the specific surface area of aggregate (m^2/kg), G is the percentage retained on sieve 9.5 mm, g is the percentage passing on sieve #9.5 mm e retained on sieve 4.8 mm, A is the percentage passing on sieve #4.8 mm e retained on sieve 0.3 mm, a is the percentage passing on sieve #0.3 mm e retained on sieve 0.074 mm and f is the percentage passing on sieve 0.074 mm.

Then, the probable OAC was calculated, using the following expression:

$$T_{ca} = m\sqrt[5]{S} \tag{2}$$

where T_{ca} is the OAC in relation to the mass of the aggregates (%) and m is the richness modulus of AC, varying from 3.75 (wearing course with high stiffness) to 4.00 (wearing course with low stiffness).

If the mean bulk specific gravity of the total aggregate is less than 2.60 or greater than 2.70, then the content obtained in the previous item should be corrected by the following expression:

$$T'_{ca} = \frac{2,65\,T_{ca}}{\delta_{am}} \tag{3}$$

where T'_{ca} is the corrected OAC in relation to the mass of the aggregates (%) and δ_{am} is the mean bulk specific gravity of the total aggregate.

Finally, the OAC is calculated in relation to the entire mixture:

$$P_{ca} = \frac{100\,T_{ca}}{100 + T_{ca}} \text{ or } P_{ca} = \frac{100\,T'_{ca}}{100 + T'_{ca}} \tag{4}$$

where P_{ca} is the final value of OAC in relation to the total mixture (%).

From that OAC value were adopted two points below it (each 0.5%) and two points above it (each 0.5%).

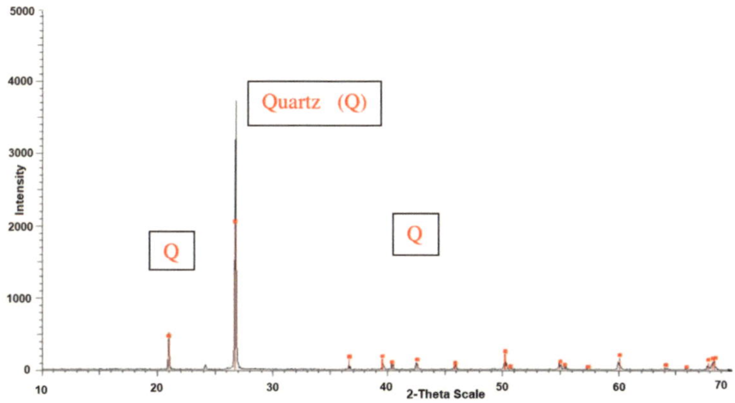

Figure 2.
XRD analysis for WFS filler.

3.4 Production of SMA samples in the laboratory

Five HMAC mixtures were analysed whose grain size proportions are shown in **Table 4**. The mixture 1 was used as reference, for 100% of Portland cement as mineral filler. The other mixtures used WFS as mineral filler, replacing Portland cement in gradual proportions each 25%. At the end, the results were compared between the mixtures with and without WFS according to the physical and mechanical tests performed.

The experimental procedures were defined as follows, for each mixture [33]: (i) determination of the AC working temperatures from Saybolt-Furol viscosity test in the range of 85 ± 10 and 140 ± 15 SSF for mixing and compaction, respectively; (ii) the components (aggregates + AC) were mixed at a temperature of 146°C for approximately 2 min; (iii) the mix was placed in the Marshall mould and compacted mechanically with 75 blows on each side of the specimen; (iv) the specimen were left at rest for 24 h at room temperature; (v) after that, the specimens were left in a water bath at 60°C for 2 h; (vi) finally, they were placed in the compression mould and submitted to compression in order to determine the rupture load and flow value. Thus, all physical and mechanical parameters of HMAC mixtures were determined by the Marshall method.

From Eq. 4, an initial OAC value of 6.15% for mixture 1 was adopted, with m = 3.75. Nevertheless, the mixture showed excessive fluid, with AC in excess. Hence, OAC = 4.5% was considered. It is noteworthy that three specimens were cast for each AC content to find the final OAC of each the mixture (mixtures 1–5), whose range varied from 3.5 to 5.5%, at each interval of 0.5%. **Figure 3a** presents the results of OAC for each mixture.

3.5 Physical and mechanical properties of SMA mixtures

After the tests, the Marshall parameters of the mixtures were determined: bulk specific gravity (BSG), theoretical maximum specific gravity (TMG), air void volume (AVV), voids in the mineral aggregate (VMA), voids filled with asphalt

Oxide	SiO_2	Al_2O_3	SO_3	Fe_2O_3
Content (%)	93.68	3.97	1.66	0.41

Table 4.
Composition of oxides present in WFS filler.

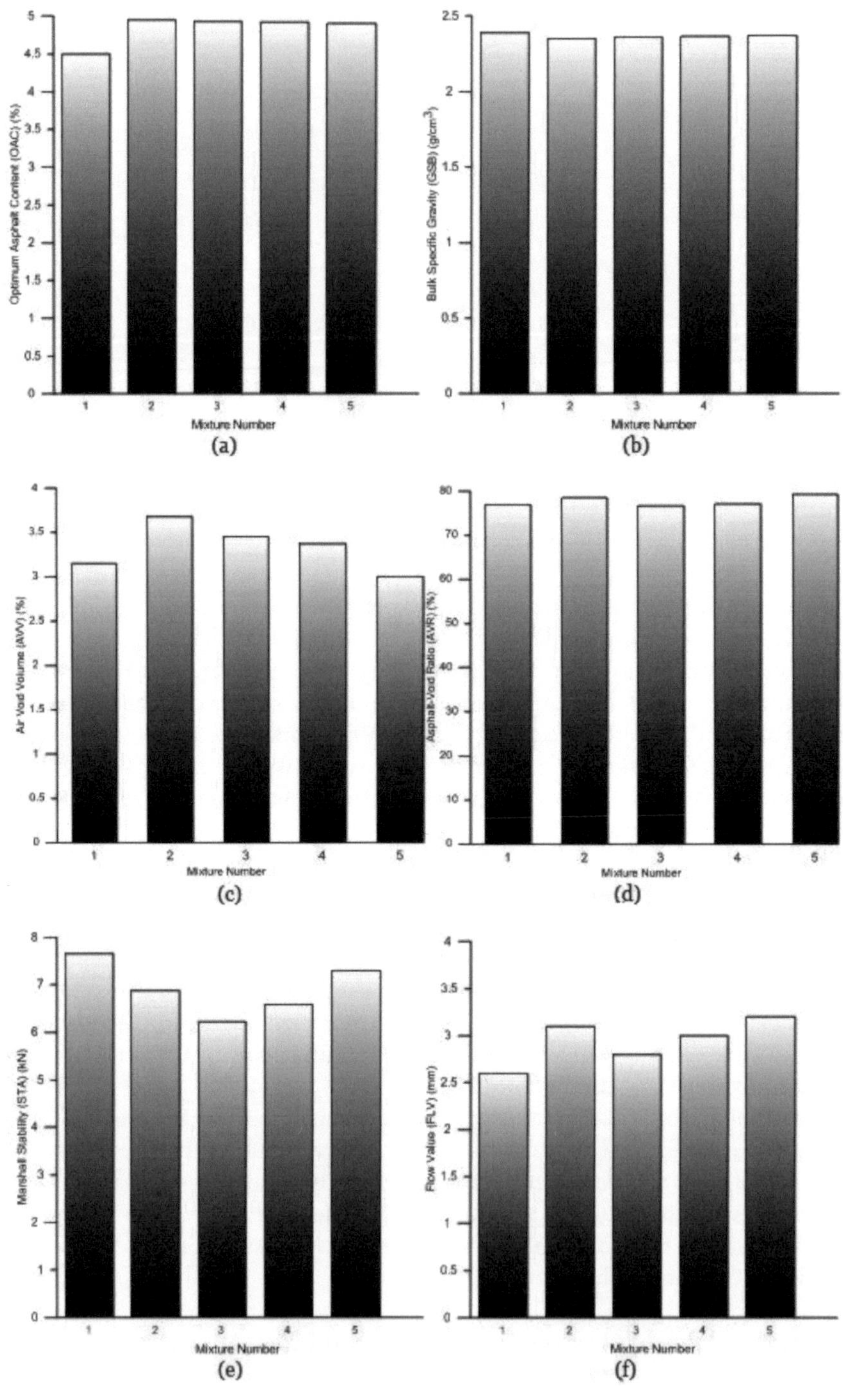

Figure 3.
Marshall physical and mechanical characteristics of studied mixtures: (a) optimum asphalt content, (b) bulk specific gravity, (c) air void volume, (d) asphalt-void ratio, (e) Marshall stability and (f) flow value.

(VFA), asphalt-void ratio (AVR), Marshall stability (STA) and flow value (FLV). The optimum contents of AC adopted were those with an AVV value of 4%.

Three samples with cylindrical forms were moulded for the determination of the static indirect tensile strength (ITS) by diametrical compression for each type of mixture, at each OAC. The ITS individual value was obtained through the expression

$$\sigma_t = \frac{T}{\pi rh} \tag{5}$$

where σ_t is the individual static ITS (kPa), T is the static rupture load (kN), *r* is the sample radius (m) and h is the sample height (m).

Three samples were moulded for determining the resilient modulus (RM) of each mixture. This mixture was then placed in the mould and compacted mechanically with 75 blows on each side of the sample. Then, the specimen were submitted to a repeatedly vertical compression load F at a maximum stress level less than or equal to 20% of the ITS. The RM adopted was the arithmetical mean value determined at 300, 400 and 500 load application F.

Hence, the value of the RM was determined by the expression [33]

$$RM = \frac{F}{\delta h} \times (0,9976\mu + 0,2692) \tag{6}$$

where RM is the individual resilient modulus (MPa), F is the cyclic vertical load diametrically applied on specimen (N), δ is the elastic strain recorded for 200, 400 and 500 load applications (mm), h is the sample height (mm) and μ is Poisson's ratio.

The fatigue test was performed to define the number of loading repetitions as a function of controlled stresses in diametrical compression samples with the load applied at a frequency of 1 Hz, with 0.10 s of repeated loading duration through the same resilient modulus equipment, increasing in tensile strain until the specimen is completely disrupted at a constant temperature of 25°C. The fatigue curve was determined in seven stress levels (7.5, 10, 15, 20, 25, 30 and 40% of the static ITS) with two specimens per level. The fatigue resistance was evaluated according to the fatigue curves generated by testing, which introduces the relationship between fatigue strength and fatigue life. The fatigue equation in this study was calculated using the formula given in the following equation [34]:

$$\log(N_f) = n \times log(\sigma_f) + k \tag{7}$$

where N_f is the fatigue life (in cycles) and σ_f is the fatigue stress (MPa), i.e. the tension stress applied during the test. The equation provides a linear relationship between them using a bilogarithmic scale, in which "n" is the gradient and "k" is the intercept.

The study of permanent deformation was made using the static creep test applying a static and continuous compression load on a specimen moulded according to the Marshall methodology. The specimen was placed in the axial position and then was subjected to an applied tension of 0.1 MPa, distributed over the entire contact surface of the specimen for a period of 60 min at a temperature of 40°C. The permanent deformations were measured continuously along that time, and then the specimen was discharged, waiting for 15 min for the stabilisation of the viscous deformations, which were measured continuously too. The total strain (D_t) after the recovery period can be obtained as:

$$D_t = \frac{\Delta h_{75}}{h_o} \tag{8}$$

where Δh_{75} is the specimen height change after the final recovery period, i.e. 75 min after the start of the test load (mm), and h_o is the specimen initial height taken in the axial direction of loading (mm). **Table 5** shows the mechanical tests performed on HMAC mixtures, while **Figure 4** presents all tests carried out on components and mixtures.

Brazilian standard	Title	Acceptance parameter (Brazilian standard)	Similar international standard
DNER-ME 043/1995	Asphalt mixtures—Marshall test	OAC ≥ 6% STA ≥ 5 kN 3% < AVV < 5%	ASTM D5581-07a
NBR 16018/2011	Asphalt mixture—stiffness determination by repeated load indirect tension test	—	ASTM D4123–82
NBR 15087/2012	Asphalt mixtures—determination of tensile strength by diametrical compression	≥0.65 MPa	ASTM D 6931–17
DNER-ME (provisional standard)/2017	Hot-mixed asphalt concrete—fatigue under repeated loading, constant tension, using the indirect tension test	—	FHWA-Protocol P07/2001
—	Standard test methods for tensile, compressive and flexural creep and creep rupture of plastics	$D_t \leq 0.02$ mm/mm in 75 min	ASTM D 2990–09

Table 5.
Mechanical characterisation tests carried out on HMAC mixtures.

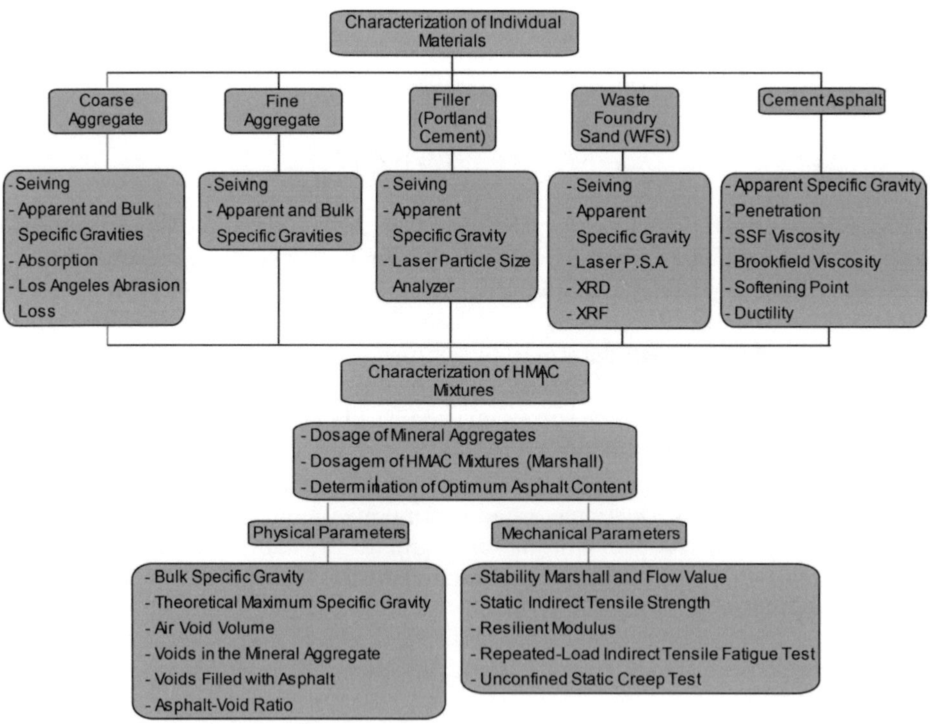

Figure 4.
Flowchart of the laboratory tests.

4. Results and discussion

4.1 Characterisation of materials

Figure 2 indicates the result of XRD analysis for WFS filler. As shown in the figure, WFS is essentially formed by quartz mineral, as expected. **Table 4** shows the composition of the main oxides present in the WFS filler obtained by XRF analysis. The high percentage of silica confirms the XRD analysis of the material [8, 16, 17]. **Table 6** indicates the physical characteristics of the aggregates. WFS aggregate apparent specific gravity of WFS is very close to conventional aggregates (pebble and sand) [7, 9, 10, 13] each other except for PC. Pebble had a Los Angeles abrasion loss below the maximum allowed by the Brazilian standard, which is 50%. The WFS had 76.25% of its particle sizes passing at #200 sieve and are slightly larger than that of Portland cement, i.e. it is too fine to replace part of the fine aggregate of the asphalt mixes [8, 10, 12], thus demonstrating that the residue could only replace part or total filler fraction.

Table 7 shows the resulting granulometric composition of the mineral aggregates with and without the addition of WFS. It is observed that all the mixtures were composed with the same amount of aggregates, varying only the proportion between the two types of the filler fraction. Conventional mixture 1 used PC exclusively, while mixture 2 used WFS as filler exclusively. The other mixtures had variations between permutations of PC and WFS proportions. The grain size distribution of the mineral aggregates, the "C" range maximum and minimum limits of the Brazilian highway specification and the resulting aggregates of mixtures 1 and 2 are shown in **Figure 5**.

Aggregate	Apparent specific gravity (g/cm^3)	Absorption (%)	Los Angeles abrasion loss (%)	d_{90} (mm)	d_{50} (mm)	d_{10} (mm)
Pebble	2.66	1.92	40.0	12.0	7.0	2.5
Sand	2.63	—	—	1.5	0.35	0.12
Filler (PC)	3.03	—	—	0.063	0.020	0.004
Filler (WFS)	2.65	—	—	0.133	0.040	0.004

Notes: d_{90}, d_{50} and d_{10} are the particle size for which 90, 50 and 10% of the all particles, in mass, are finer than it.

Table 6.
Physical characteristics of aggregates.

Aggregate	Mixture designation				
	1 (%)	2 (%)	3 (%)	4 (%)	5 (%)
Pebble	62.0	62.0	62.0	62.0	62.0
Sand	33.0	33.0	33.0	33.0	33.0
Filler (cement)	5.0	0.0	3.75	2.5	1.25
Filler (WFS)	0.0	5.0	1.250	2.5	3.75
% Total	100.0	100.0	100.0	100.0	100.0

Table 7.
Granulometric composition of mineral aggregate mixtures with and without WFS addition.

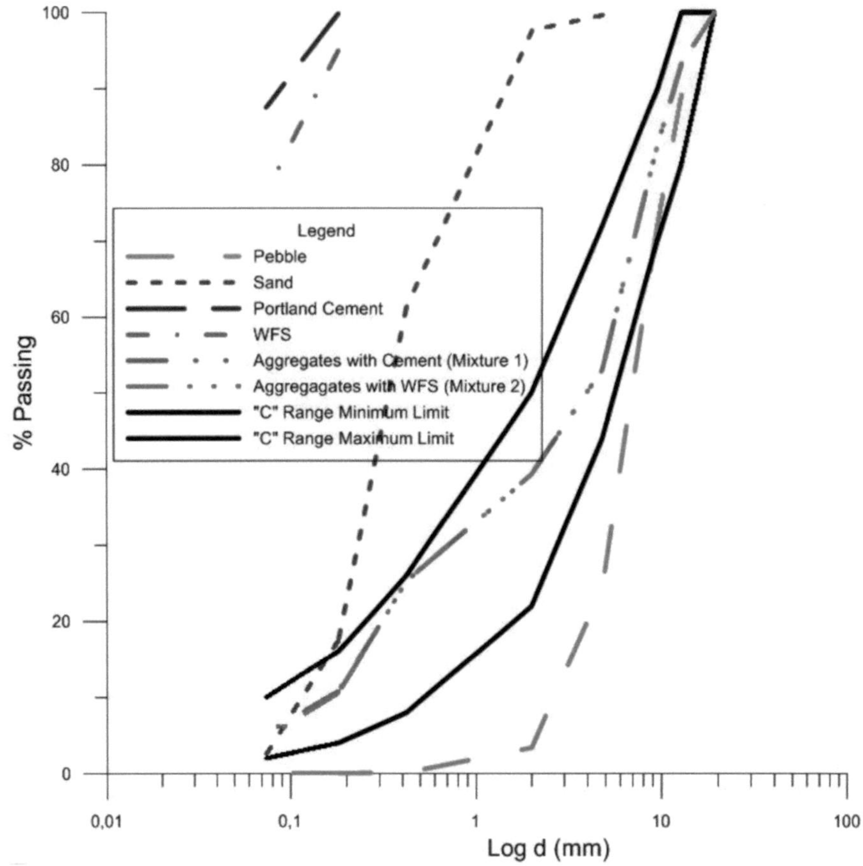

Figure 5.
Grain size distribution and limit curves of mineral aggregates.

4.2 Physical characteristics of mixtures

Figure 3 shows the main physical parameters of the mixtures, obtained through the Marshall methodology. OAC values of the mixtures containing WFS are comparable to the control in mixture 1 [14, 17]. Mixture 1 obtained the lowest OAC (4.5%), whereas mixtures with WFS had little bit higher OAC values, whose contents increased as WFS proportions were increased too [13]. This reason probably is due to the absorption characteristics of this residue, and not due to the grain size [10, 32], since CP has larger particle size and therefore smaller surface area and thus should consume less AC, at the same proportion of WFS.

It was observed that all five mixtures met the Brazilian standards regarding the physical Marshall parameters (OAC, AVV, VMA and AVR). Mixture 1 had a higher GMB values than all other mixtures with WFS and was therefore the densest. The other mixtures maintained a slight decrease of this parameter, when the proportion of WFS in the mixture was increased [7, 9, 13]. Mixture 2 (100% WFS filler) had the highest amount of AVV and the second largest AVR among all mixtures. AVV values increased when WFS content were increased in the mixtures [8–10, 32].

4.3 Mechanical characteristics of mixtures

High amounts of AVV and AVR tend to negatively influence STA and FLV values, given the viscous characteristic of AC. Thus, mixture 1 showed the best performance, with the highest STA and lowest FLV values. Among the mixtures using WFS, mixture 5 (one fourth WFS + three fourths PC) was the one that presented the highest value of GMB, thus being the densest, and also presented the highest value of STA; however, it had the highest FLV value too. The FLV values of the WFS blends were higher than the PC blends, which characterises a higher AC consumption of these blends. In summary, the use of WFS decreased the stability of blends [6, 8, 10, 12, 32] while increasing their fluency. This latter is in disagreement with that observed by the author cited previously. Even so, all mixtures showed STA values higher than the minimum required (>5 kN).

There was a certain tendency that static ITS values will decrease as WFS content increased [7, 9, 10, 13, 32]. Mixtures 3 and 5 presented higher values of this parameter than control mixture 1 [25]. All asphalt mixtures presented values above the minimum value of the Brazilian standard (>0.65 MPa). This is a good indication for durability of the mixtures since fatigue life is a function of ITS. There was not an apparent correlation between AVV and static ITS values (**Figure 6**).

The use of WFS decreased the RM values. Mixture 1 presented the highest value, followed by mixture 2. In Brazil, the relationship between RM and static ITS (RM/ITS) has been used as an analysis parameter to evaluate the behaviour of asphalt mixtures related to fatigue life. As a rule, mixtures with RM/ITS ratio around 3000

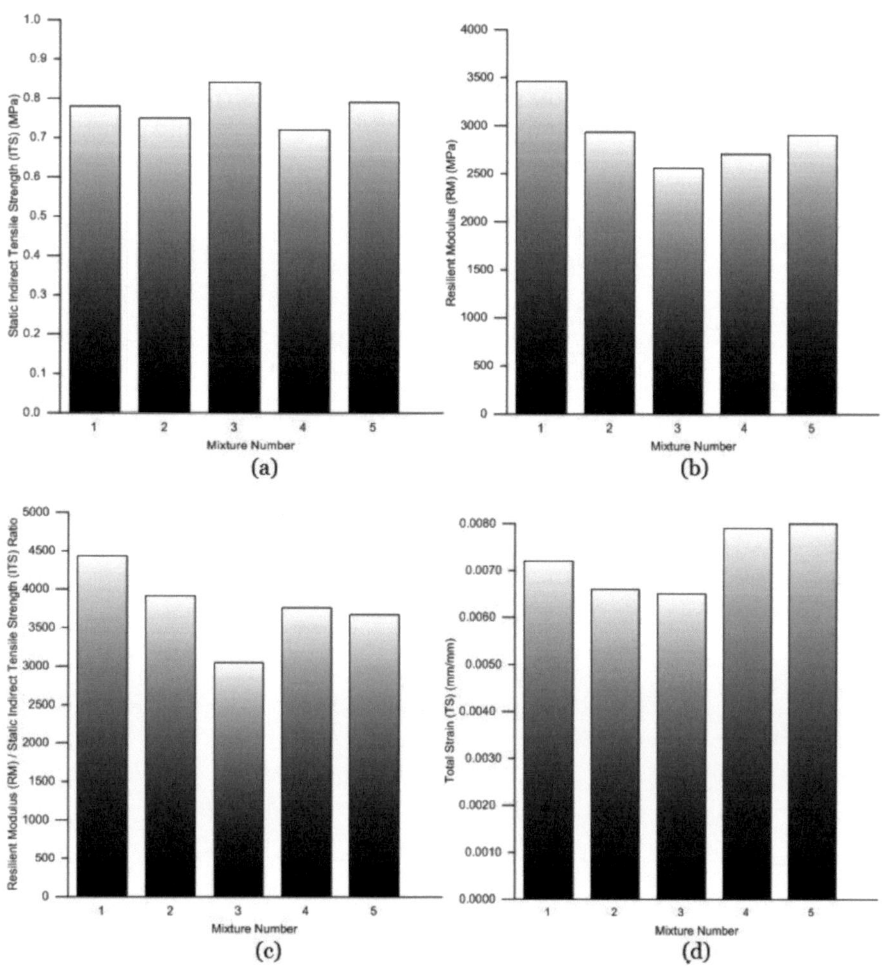

Figure 6.
Mechanical characteristics of studied mixtures: (a) static indirect tensile strength, (b) resilient modulus, (c) RM/ITS ratio and (d) total strain (static creep).

exhibit good structural behaviour because they allow the use of thinner asphalt wearing layers for the same fatigue life; that is, they characterise mixtures that are not susceptible to early development of permanent deformations because they are not rigid enough. In this sense, mixture 3 was the only one that met this criterion. On the other hand, the conventional mixture 1 presented the highest value of this ratio, thus indicating a more rigid behaviour.

Figure 7 shows the comparison between asphalt mixtures in relation to the stress-controlled fatigue test. For the acquisition of fatigue curves, the average value of the RM and the static ITS of each mixture were used. Between Mixtures 1 and 4, the best-fitting straight lines were very close to each other, with a parallelism between the line slopes, and both mixtures can be considered to have practically the same fatigue life. Mixture 2 presented the shortest fatigue life, while mixture 5 presented the longest fatigue life, standing out among the others. For applied stress differences up to 0.4 MPa, Mixtures 1, 2 and 4 behave similarly.

It should be noted that mixture 5 presented the second best ITS result and the second closest value of the RM/ITS ratio around 3000, thus justifying the use of this parameter as a quantitative indicator of fatigue life of asphalt mixtures. The fatigue life test on mixture 3 was not performed.

Regarding the permanent deformation, Mixtures 2 and 3 presented lower values than mixture 1, while mixture 5 presented the highest value among the others. There was no direct relationship with AVV, since, of all of them, mixture 5 presented the lowest value of voids. Mixture 3 presented the lowest value of permanent deformation, confirming again the good indicative of the RM/ITS ratio around 3000 in predicting the behaviour of asphalt mixtures for fatigue and permanent deformations. All mixtures presented permanent deformation values below the

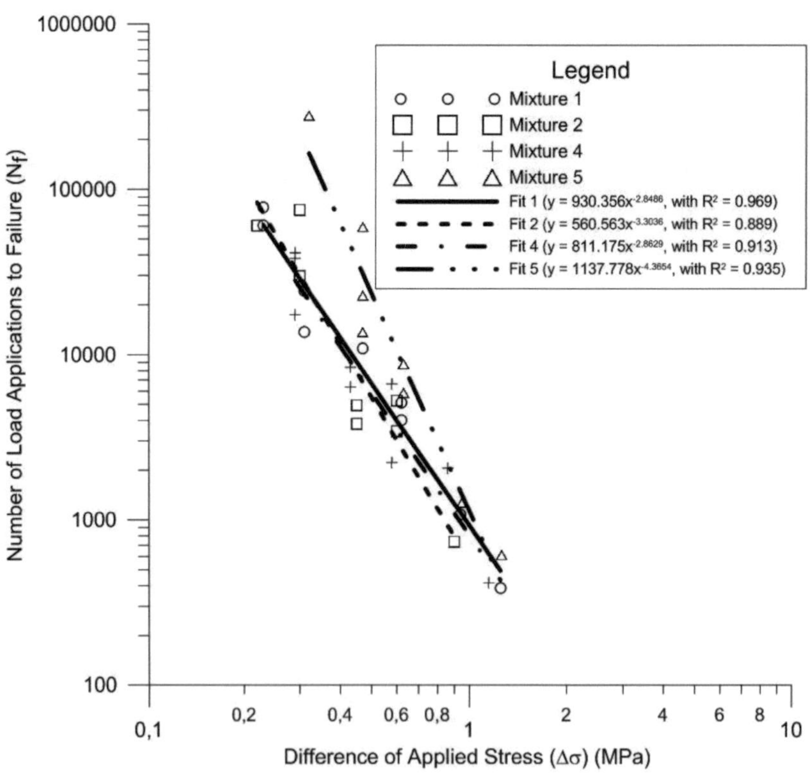

Figure 7.
Fatigue life for the mixtures with and without WFS content.

conventional criterion of 0.020 mm/mm and do not have the tendency to be susceptible to premature permanent deformations.

5. Conclusions

This work analysed five asphalt mixtures, one using 100% CP as a filler and the other four using WFS, with a maximum proportion of 5% (by weight) of the total aggregate. The WFS residue used consisted of almost 94% silica, without organic compounds, with apparent specific gravity similar to clean sand and slightly coarser than CP.

All mixtures with WFS residue presented physical and mechanical parameters within the Brazilian standards, following the Marshall methodology, although with lower STA and higher FLV values. The use of WFS increased static ITS values, while decreased MR values. The mixtures with WFS showed total permanent deformation values less than 2% after 75 min of the test. The RM/ITS ratio around 3000 proved to be a good indication of mixtures with better performance against fatigue life and permanent deformation.

Finally, the use of WFS as a mineral filer in asphalt mixtures proved to be adequate, meeting the criteria of Brazilian standards in physical and mechanical tests.

Acknowledgements

The authors would like to thank Prof. Dr. Laura Maria Goretti da Mota, from Pavement Laboratory of COPPE/Federal University of Rio de Janeiro (UFRJ), for some laboratorial tests carried out in that place. This work was supported by the CNPq [grant number 620244/2008-9]; FAPEAM [scholarship].

Author details

Nilton de Souza Campelo[1*], Karine Jussara Sá da Costa[2], Raimundo Kennedy Vieira[1] and Adalena Kennedy Vieira[1]

1 Master Science Program in Civil Engineering, Federal University of Amazonas, Manaus, AM, Brazil

2 Department of Civil Engineering, Federal University of Roraima, Boa Vista, RR, Brazil

*Address all correspondence to: ncampelo@ufam.edu.br

References

[1] Abichou T, Edil TB, Benson CH, Bahia H. Beneficial use of foundry by-products in highway construction. In: Proceedings of Geotechnical Engineering for Transportation Projects (GeoTrans2004); 27-31 July 2004; Los Angeles, California, USA; 2004. pp. 715-722

[2] Abichou TH, Edil T, Benson CH, Tawfiq K. Hydraulic conductivity of foundry sands and their use as hydraulic barriers. In: Recycled Materials in Geotechnics Sessions at ASCE Civil Engineering Conference and Exposition 2004. Vol. 1. Baltimore, Maryland, United States; 19-21 October 2004. pp. 186-200. DOI: 10.1061/40756(149)13

[3] Javed S, Lovell CW. Use of foundry sand in highway construction. In: Proceeding of the 44th Highway Geology Symposium. Tampa, Florida, The Mayan; 1993. pp. 19-21, 19-34

[4] Bolden J, Abu-Lebdeh T, Fini E. Utilization of recycled and waste materials in various construction applications. American Journal of Environmental Science. 2013;**9**(1): 14-24. DOI: 10.3844/ajessp.2013.14.24

[5] Bradshaw SL, Benson CH, Olenbush EH, Melton JS. In: Proceedings of the Green Streets and Highways Conference 2010. Vol. 1. Denver, Colorado, USA: ASCE; 14-17 November 2010. pp. 280-298

[6] Dyer PPOL, Lima MG, Klinsky LMG, Silva SA, Coppio GJL. Environmental characterization of foundry waste sand (WFS) in hot mix asphalt (HMA) mixtures. Construction and Building Materials. 2018;**171**:474-484. DOI: 10.1016/j.conbuildmat.2018.03.151

[7] Bakis R, Koyuncu H, Demirbas A. An investigation of waste foundry sand in asphalt concrete mixtures. Waste Management and Research. 2006;**24**:269-274. DOI: 10.1177/0734242X06064822

[8] Javed S. Use of Waste Foundry Sand in Highway Construction (Final Report). School of Civil Engineering, Purdue University, West Lafayette, Ind., Project No. C-36-50N, Report JHRP-94/2; May 1994

[9] Javed S, Lovell CW, Wood LE. Waste foundry sand in asphalt concrete. Transportation research record 1437. Transportation Research Board. Washington, DC; 1994:27-34

[10] Pittenger DM. State-of-the-practice literature scan for foundry sand. Gallogly College of Engineering, University of Oklahoma, Oklahoma Department of Transportation; 2017

[11] EPA. Beneficial Reuse of Foundry Sand: A Review of State Practices and Regulations. Washington, D.C: U.S. Environmental Protection Agency; 2002

[12] Gedik A, Lav AH, Lav MA. Investigation of alternative ways for recycling waste foundry sand: An extensive review to present benefits. Canadian Journal of Civil Engineering; 2017:1-36

[13] Dyer PPOL, Lima MG, Klinsky LMG, Silva SA, Coppio GJL. Macro and microstructural characterisation of waste foundry sand reused as aggregate. Road Materials and Pavement Design. 2019:1-14 (published online). DOI: 10.1080/14680629.2019.1625807

[14] Miller E, Bahia HU, Benson C, Khatri A, Braham A. Utilization of Waste Foundry Sand in Hot Mix Asphalt Mixtures. University of Wisconsin, Madison, Wisconsin: American Foundry Society; 2001. pp. 01-103

[15] CWC. Beneficial reuse of spent foundry sand. In: Clean Washington Center, IBP-95-1. Seattle, Washington; 1996. Available from: www.cwc.org.br

[16] Pasetto M, Baldo N. Laboratory investigation on foamed bitumen bound mixtures made with steel slag, foundry sand, bottom ash and reclaimed asphalt pavement: Cold recycling and bitumen stabilization technology. Road Materials and Pavement Design. 2012;**13**(4):691-712. DOI: 10.1080/14680629.2012.742629

[17] Tikalsky PJ, Bahia HU, Deng A, Snyder T. Excess foundry sand characterization and experimental investigation in controlled low-strength material and hot-mixing asphalt (Final Report). The Pennsylvania State University, Transportation Research Building, University Park, PA, U.S.A., U.S. Department of Energy, Contract No. DE-FC36-01ID13974; 2004

[18] American Foundrymen's Society. Alternative utilization of foundry waste sand. Final Report (Phase I) prepared by American Foundrymen's Society Inc. For Illinois Department of Commerce and Community Affairs, Des Plaines, Illinois; 1999

[19] Guney Y, Aydilek AH, Demirkan MM. Geoenvironmental behavior of foundry sand amended mixtures for highway subbases. Waste Management. 2006;**26**:932-945. DOI: 10.1016/j.wasman.2005.06.007

[20] Kleven JR, Edil TB, Benson CH. Evaluation of excess foundry system sands for use as subbase material. In: Proceedings of the 79th Annual Meeting, Transportation Research Board. Washington, DC: CD-Rom; 2000. 27 p

[21] Yazoghli-Marzouk O, Vulcano-greullet N, Cantegrit L, Friteyre L, Jullien A. Recycling foundry sand in road construction–field assessment. Construction and Building Materials. 2014;**61**:69-78. DOI: 10.1016/j.conbuildmat.2014.02.055

[22] Fox PJ, Mast DG. Geotechn1cal performance of highway embankment constructed using waste foundry sand (Final Report). School of Civil Engineering, Purdue University, West Lafayette, Ind., Project No. C-36-36Z, FHWA/IN/JTRP-98/1; 1998

[23] Partridge BK, Fox PJ, Alleman JE, Mast DG. Field demonstration of highway embankment construction using waste foundry sand. Transportation research record 1670. Transportation Research Board. Washington, DC; 1999;**1670**(1):98-105. DOI: 10.3141/1670-13

[24] Abichou T, Benson C, Edil T. Foundry green sand as hydraulic barriers: Field study. Journal of Geotechnical and Geoenvironmental Engineering. 2002;**128**(3):206-215

[25] Abdulsattar ZA, Mohammed EA. Effect of waste foundry sand on indirect tensile strength of asphalt mixture. Journal of Engineering and Sustainable Development. 2018;**22**(4):116-123. DOI: 10.31272/jeasd.2018.4.9

[26] Coutinho B, Furlan AP, Fabbri GTP. Evaluation of the reuse of foundry sand as aggregate in dense asphalt mixtures. In: Proceedings of the International Symposium on Pavement Recycling (ISPR '05), ISPR; 14-16 March 2005; São Paulo/SP, Brazil; 2005

[27] Winkler ES, Bol'shakov AA. Characterization of foundry sand waste. Center for Energy Efficiency and Renewable Energy, University of Massachusetts at Amherst, Chelsea Center for Recycling and Economic Development, Technical Research Program; 2000

[28] Shah V, Aijaz A, Naidu K, Francis S, Kayam PK. Waste aggregate in concrete pavement—A review. International

Journal of Latest Engineering Research and Applications (IJLERA). 2016;**1**(9):44-50

[29] Abichou T, Benson CH, Edil TB. Beneficial reuse of foundry by-products. In: Environmental Geotechnical Report 99-1. Madison, Wisconsin, USA: Department of Civil and Environmental Engineering, University of Wisconsin-Madison; 1999

[30] Chaudhari AG, Patekar PS, Khan JK, Rokade VJ, Siddiqui AS. Review study: On replacement of fine aggregate by foundry sand in a pavement. International Journal of Innovative Research in Science and Engineering. 2017;**3**(3):402-405

[31] FHWA. User Guidelines for Waste and Byproduct Materials in Pavement Construction, Publication Number: FHWA-RD-97-148. Available from: https://www.fhwa.dot.gov/publications/research/infrastructure/structures/97148/intro.cfm

[32] Suji D, Poovendran S, Prabhakar P. Experimental study on partial replacement of waste foundry sand in flexible pavements. International Journal of Civil and Structural Engineering Research. 2016;**4**(1):188-197

[33] Campelo NS, Campos AMLS, Aragão AF. Comparative analysis of asphalt concrete mixtures employing pebbles and synthetic coarse aggregate of calcined clay in the Amazon region. International Journal of Pavement Engineering. 2017;**20**(5):507-518. DOI: 10.1080/10298436.2017.1309199

[34] Liu Y, Han S, Zhang Z, Xu O. Design and evaluation of gap-graded asphalt rubber mixtures. Materials and Design. Elsevier Ltd; 2012;**35**:873-877. DOI: 10.1016/j.matdes.2011.08.047

Section 2

Environmental Impacts and Sustainability

Chapter 4

Environmental Impact and Sustainability of Aggregate Production in Ethiopia

Gashaw Assefa and Aklilu Gebregziabher

Abstract

The production of aggregate for the infrastructural development of the country has been increasing for the last three decades due to the high urbanization rates in the main cities of the country and the ever-growing demand for basic infrastructural facilities. The environmental impact of both fine and coarse aggregate production is now hard to ignore especially on the outskirts of the main cities. These impacts are clearly seen on the degradation of landscape and land stability, pollution of water resource, pollution of the atmosphere due to dust, and societal impacts. There are clear local and international laws that protect the environment from the negative impact of any project, whereas the observed fact from abandoned and functioning quarry sites shows these rules are not followed strictly.

Keywords: aggregate, sustainability, Ethiopia, production, environmental impact

1. Introduction

The construction industry in Ethiopia is a major driving sector for economic growth. Based on a report by the National Bank of Ethiopia [1], the construction industry in 2018 accounted for 71.4% of the nations' industrial output and expanded by 15.7% from its previous share signifying the leading role of construction sector. Massive government investment in infrastructure and residential building projects has made the sector to create jobs and improve standard of living. The rate of urbanization in the major cities is increasing, and this created a huge need for improved infrastructure systems and a big housing project. [2] reported that the urbanization rate between 1984 and 2007 has quadrupled from 3.7% to 14% over the two decades and still it was one of the lowest in the world, well below the sub-Saharan African average of 37%; it is projected that in 2028 the 30% of the country people live in urban areas [3].

Coarse and fine aggregates are the major inputs for most of the infrastructural systems and building projects. Approximately three quarter of the volume of concrete is occupied by aggregate and 90% of asphalt pavement is aggregate [4, 5]. This increased the demand of aggregate in line with the sectoral development. Yasmin, in 2015 based on the annual cement production volume, estimated the demand of Ethiopian for sand would be approximately 1.5 million tons per year [6].

2. Sourcing and aggregate production for Ethiopian construction industry

2.1 Sources of aggregate in Ethiopian past, present, and future trends

The natural aggregates are formed as a result of the processes of weathering and abrasion or through crushing a large parent mass [7]. Engineers are first of all concerned with technical requirements. However, in the future, probably the environmentalists will take over much of the standardization work. Quarrying and transport of materials have environmental impacts on the local neighborhood and society, for instance, with regard to noise, dust, pollution, and effects on biodiversity [8]. The city of Addis Ababa is growing from time to time very rapidly. Its area which is 54,000 Ha is being covered by buildings, houses, roads, bridges, etc. [9]. To meet the overgrowing demand, the number and production capacity of quarry sites, coarse aggregate, and sand deposits are aggressively increasing. Admasu [7] (2015) reported that based on the data from Addis Ababa Environmental Protection Authority in 2005 to 2015, the number of aggregate production plants in the capital has increased from 152 to 257.

2.2 Production of fine aggregate type, quality, production, and quarrying trend and method

For central and southeastern part of the country, the most common sources of sands are from Meki, Langano, Sodere, Koka, Metehara, and Minjar [10]. Generally, the existing quarry nationwide are not well organized; for example, from all sites in the capital, Denamo [9] reported, only 10 of the existing aggregate producing firms are well organized in manpower, machinery, and finance. Gravel and stone quarry operations result in extensive manipulation of the landscape and of the ecosystems of indigenous to their sites [11] (**Figure 1**).

Figure 1.
Sources of sands for central and southeastern part of the country [source: Google Maps 2019].

Figure 2.
Production of fine aggregate in Ethiopia [6].

Properties	Result	Properties	Result
Dynamic elasticity modulus (GPa)	64–129	Bulk density (g/cm^3)	2.6–3.1
Ultrasonic pulse velocity (m/s)	4000–7000	Compressive strength (MPa)	130–350
Open porosity (%)	0.33%–3.08		

Table 1.
Physical and mechanical properties of Termaber basalt [15].

Quarrying activity often has long-term social and environmental impacts. Social challenges related to the increase in quarrying activities in general include threats to health and safety, farming obstacle, blockage on free movement of animals, displacement of communities, reduction in agricultural yield, damage of cultural sites, and the formation of mining villages [12]. Production activities in any industry may harm the environment through their damaging effects on air, water, soil, and biodiversity [13]. Due to the production process and luck of standardization, there is a big problem in getting good sand for production of concrete due to different reasons (**Figure 2**).

2.3 Production of coarse aggregate type, quality, production, and quarrying trend and method

Production of coarse aggregates includes blasting of rock, transporting of the crushed rock by conveyor to the crushing plant, and adjusting the crusher so as to give a range of different sizes by passing the crushed rock through a set of sieves [7]. At the selected quarry sites, holes are drilled into the rock and are partially filled with explosives, and controlled sequential blasting commonly breaks the rock into pieces suitable for crushing. If the rubble is too large, secondary breaking may be required and usually is accomplished with hydraulic hammers [14].

A detailed study was conducted by Tesfaye A. and Giulio B. on Termaber basalt, a widely used basalt in the central highland of Ethiopia and that comprise the major source of local crushed rock aggregates and building stone. Based on field investigation and laboratory tests, it was concluded that the basalt was highly suitable to be used as construction material with listed properties in **Table 1** [15].

3. Environmental impact of aggregate production

Production activities in any industry may harm the environment through their damaging effects on air, water, soil, and biodiversity [6]. Sustainable supply of aggregate mix goes beyond the need to ensure a secure supply of aggregates to the economy by adding the requirement that the selected blend of natural aggregates, quarry by-products, and recycled waste must be produced and transported in an eco-efficient manner that minimizes total negative impacts and maximizes overall benefits to society [16]. Locally there are different environmental impacts due to the quarry sites of aggregates; the major impacts are summarized in the following section.

3.1 Impact on landscape and land stability

Stripping, excavating, damping of the overburden soil, drilling, and blasting of the rocks are the serious causes of soil erosion due to the quarry sites. Due to this activity, frequent soil creeps, siltation of down steams, formation of pits, borehole, surface ragging or cliff are created. Besides, agriculture productivity redaction, losses of the natural aesthetic of the area, and the bio diversity are created [14] (**Figures 3** and **4**).

3.2 Impact on the atmosphere

When a blast is detonated, some energy will escape into the atmosphere causing a disturbance in the air. Part of this disturbance is subaudible (air concussion) and part can be heard (noise). Once again the same to noise effect, the repetition and the exposure of the workers repeatedly make potential causes for the airborne diseases, respiratory infection, etc. [14].

3.3 Impact on water resource

Groundwater flow in springs, gaining streams, and wells may be impacted by nearby aggregate operations that pump groundwater from the pit or quarry [9]. Due to resuspension of sediments, sedimentation due to stockpiling and dumping of excess mining materials and organic particulate matter, and oil spills or leakage from excavation machinery and transportation vehicles, increased turbidity in sand mining sites are common [6] (**Figure 5**).

3.4 Impacts on biodiversity

It has been reported that due to fine aggregate extraction, many hectares of fertile streamside land are lost annually, as well as valuable timber resources and wildlife habitats in the riparian areas. Degraded stream habitats result in loss of

Figure 3.
(A) Degradation of quarries in Addis Ababa, around Debre Gelan [14] and (B) Scoria Source, Tullu Dimtu, around Debre Gelan [14].

Figure 4.
Abandoned quarry site at Augusta, Addis Ababa, Ethiopia [11].

Figure 5.
Water pollution by liquid waste discharge from one of the biggest quarry of Midroc, Addis Ababa [6].

Figure 6.
Coarse aggregate crushing plants have effect on nearby inhabitants [14].

fisheries productivity, biodiversity, and recreational potential. Severely degraded channels may lower land and aesthetic values [6].

3.5 Social impacts of quarry

If the proposed quarry is in urban center where it is surrounded by residential and recreational land of high scenic values, quarry operation will negatively impact on these values. The quarry would be visible to homes, parks, and open space.

The quarry operations will produce fugitive dust from blasting, vehicular emissions, and other mining operation which would deteriorate air quality. The dust will affect negatively the health and the well-being of residents [11]. A descriptive example of such incident is demonstrated by Semere [14]. He reported in 2013 that there was a functioning coarse aggregate quarry site 60 m from a cooperative apartment dwelling in the capital city of Ethiopia; this has an impact both on the structures and the dwellers around the vicinity (**Figure 6**).

4. Legal framework

Ethiopia has signed/or ratified a number of multilateral and international agreements that aim to protect the environment. In light of these agreements, the following paragraphs summarize the local proclamations and policies to laws.

The Provisions 1995 Constitution of the Federal Democratic Republic of Ethiopia provides a perfect basic framework on which detailed laws shall be developed for various sectors. It contains provisions that support the enactment of EIA legislation. Thereof, it stipulates that the design and implementation of development programs and projects in the country should not damage or destroy the environment. It makes sure that the right of the people to be consulted and express their views on the planning and implementation of environmental policies and projects that affect them. It also states citizens have the right to live in a clean environment and, where displaced or livelihood has been adversely affected by the development projects undertaken by the government, the rights to get commensurate monetary or Overview of EIA in Ethiopia alternative compensation, including relocation with adequate state assistance [14, 17].

The 299/2002 Article 5 Environmental Impact Assessment Proclamation defines any projects that are likely to have a negative impact on the environment and requires an EIA process for any planned development project or public policy. With regard to development projects, the proclamation stipulates that no person shall commence implementation of proposed project identified by directive as requiring EIA without first passing through environmental impact assessment process and obtaining authorization from the competent environmental agency [18]. Additionally **Table 2** states the local policies and proclamations which aim to legally mitigate and control environmental impacts [19].

As per the classification of the above proclamation, aggregate quarries fails under the category of project that can likely to have a negative impact on the environment, and as per the directives given, the environmental impact assessment

Local proclamations or policies	Year
The Environment Policy of Ethiopia (April 1997)	1997
The National Policy on Biodiversity Conservation and Research (1998)	1998
Environmental Impact Assessment Proclamation 299/2002	2002
Pollution Control Proclamation 300/2002 27	2002
Prevention of Industrial Pollution: Council of Ministers Regulation No. 159/2008	2008
Environmental Organs Establishment Proclamation	

Table 2.
Local proclamations and policies related to environment protection.

study must be prepared before the aggregate production plant is established. Moreover, the proclamation imposes a fine of between 50,000 and 100,000 Birr on any project owner who commences implementation of a project without obtaining authorization from environmental agencies or who makes false presentation in the environmental impact assessment study report [18]. In addition, the proclamations clearly state that the public particularly the communities that are likely be affected by the project should comment on the environmental impact assessment study and their concern must be addressed.

5. Conclusion

The need for increased infrastructural and housing requirements due to the growing rate of urbanization has led to an ever-increasing demand for both fine and coarse aggregate. This increase in an unregulated system is raising the environment concern to a level that immediate intervention is required. Locally quarry activity has created several environmental and social problems. These problems include change of landscapes and loss of aesthetic value, contamination of soil, erosion and sedimentation, air and water pollution, loss of biodiversity, and health problems on the workers and local residents. Even if there are relatively sufficient laws and proclamations for the protection of the environment, quarry sites are not regularly inspected of by the authorized bodies while they are operational or after they have been closed for reclamation purposes. Most quarry sites are just left abandoned. The only environmental mitigation practices in few quarry sites are planting of trees and collection of waste [6, 12, 14].

Author details

Gashaw Assefa[1]* and Aklilu Gebregziabher[2]

1 Construction Technology and Management Department, Faculty of Civil Engineering and Built Environment, Hawassa University Institute of Technology, Hawassa, Ethiopia

2 Civil Engineering Department, Faculty of Civil Engineering and Built Environment, Hawassa University Institute of Technology, Hawassa, Ethiopia

*Address all correspondence to: onethiopia@gmail.com

References

[1] N. B. of E. (NBE). National Bank of Ethiopia 2017/18 annual report. 2018;**34**(1):126

[2] Schmidt E, Kedir M. Urbanization and spatial connectivity in Ethiopia: Urban growth analysis using GIS. In: ESSP II Work. Pap. Vol. 3. 2009

[3] World Bank. Ethiopia-Urbanization review: Urban institutions for a middle-income Ethiopia. 2015. [Online]. Available from: http://documents.worldbank.org/curated/en/543201468000586809/Ethiopia-Urbanization-review-urban-institutions-for-a-middle-income-Ethiopia

[4] Neville AM, Brooks JJ, Adam M. Concrete Technology. 2nd ed. England: British Library; 2010

[5] Lavin P. Asphalt Pavements: A Practical Guide to Design, Production and Maintenance for Engineers and Architects. 2003

[6] Yasmin Y, Abebe D. Fine Aggregate Production and Its Environmental Impact in Some Selected Sites of the Rift Valley Area in Ethiopia. Vol. November. Ethiopia: Addis Ababa Univ.; 2014. p. 101

[7] Admasu T. Handling of Aggregates in the Ethiopian Construction Industry: The Case of Addis Ababa. Ethiopia: Addis Ababa University; 2015

[8] Danielsen SW, Kuznetsova E. Environmental Impact and Sustainability in Aggregate Production and Use. In: Lollino G, Manconi A, Guzzetti F, Culshaw M, Bobrowsky P, Luino F, editors. Engineering Geology for Society and Territory—Vol. 5. Cham: Springer International Publishing; 2015. pp. 41-44

[9] Denamo A, Motzko C, Dinku A. Handling of Concrete Making Materials in The Ethiopian Construction Industry. Vol. October. Ethiopia: Addis Ababa University; 2005

[10] Habtamu S, Abraham A. Study on Quality of Site Concrete Production and its Management Study on Quality of Site Concrete Production and its Management Practice in Addis Ababa Housing Projects Practice in Addis Ababa Housing Projects (Case study on Koye Feche Housing Projects). Ethiopia: Addis Ababa University; 2017

[11] Enatfenta M. Impact Assessment and Restoration of Quarry Site in Urban Environment: The Case of Augusta Quarry. Vol. August. Ethiopia: Addis Ababa University; 2007

[12] Endalew A. Environment and social impacts of stone quarrying: South Western Ethiopia, in case of Bahir Dar Zuria Wereda Zenzelma Kebele. International Journal of Research in Environmental Science. 2019;**5**(2):29-38

[13] Tesfaye A. Environmental impact assessment and monitoring under Ethiopian law. Haramaya Law Review. 2012;**1**(1):103-124

[14] Mulatu S, Dinku A. Environmental Impacts of 14. 2013. p. 119

[15] Engidasew TA, Barbieri G. Geo-engineering evaluation of Termaber basalt rock mass for crushed stone aggregate and building stone from Central Ethiopia. Journal of African Earth Sciences. 2014;**99**(PA2):581-594

[16] Blengini GA et al. Life cycle assessment guidelines for the sustainable production and recycling of aggregates: the Sustainable Aggregates Resource Management project (SARMa). Journal of Cleaner Production. 2012;**27**:177-181

[17] E. Constitution. FDRE Constitution of 1995. 1994. pp. 1-40

[18] The Federal Democratic Republic of Ethiopia. Proclamation No. 299/2002 Environmental Impact Assessment Proclamation. Gov. Proclam; 2002

[19] EEPA. United Nations Conference on Sustainable Development (Rio+20) National Report of Ethiopia. Environmental Protection Authority; 2012

Section 3

Hybrid Aggregates

Chapter 5

The Influence of Hybrid Aggregates on Different Types of Concrete

Jianhui Yang

Abstract

This chapter presents different experimental results regarding the influences of normal-weight sand and lightweight sand (shale pottery (SP)) on different types of concrete. Because of the porosity of lightweight aggregates (LWAs), which can absorb and release water in concrete, the effect of concrete curing is better, and thus the properties of concrete are improved. On the other hand, because the lightweight coarse aggregate (LWCA) rises easily in all-lightweight concrete (ALWC) during pumping and vibration and the cost of ALWC is also higher, a method of replacing part of the lightweight aggregates in ALWC with normal-weight aggregates is used. These new types of concretes include sand lightweight concrete (SLWC), gravel lightweight concrete (GLWC), hybrid aggregate light-weight concrete (HALWC), and so on. This chapter mainly discusses the properties of lightweight aggregate concrete (LWAC), lightweight sand foamed concrete, lightweight sand mortar, and reinforced LWAC. The chapter also includes LWAC of high temperature, low temperature, durability, and uni- and multiaxial mechanical properties according to the results of our research group over recent decades. All of the experimental results show that the properties can meet Chinese National Code requirements.

Keywords: lightweight aggregate concrete, foamed concrete, mortar, reinforced concrete beam, durability, high temperature, low temperature, shale ceramsite, shale pottery

1. Introduction

There are notable differences between lightweight aggregates and normal-weight aggregates (NWA), so what are the different effects on different types of concrete?

Lightweight coarse aggregate has a softening effect; does lightweight aggregate concrete have this effect too?

Can the lightweight concrete be used under special environmental conditions such as negative temperature, elevated temperature, and chemical corrosion?

Are there any changes in the mechanical performance of lightweight concrete under complicated stress and in reinforced concrete?

Since 2005, it has been prohibited to mine river sand in most areas of China. The relevant laws, such as the *Water Law of the People's Republic of China* (2002, 2016),

IntechOpen

the *Flood Prevention Law of the People's Republic of China* (1997, 2016), the *Mineral Resources Law of the People's Republic of China* (1986, 2009), and the *Regulations of the People's Republic of China on the Administration of River Courses* (1986, 2009), are constantly amended, and thus the prohibition of river sand excavation has been extended throughout the country since 2018. Nowadays, artificial sand is mainly used as normal-weight fine aggregate (NWFA) in normal-weight concrete (NWC) in China. This study mainly discusses artificial sand, and all of the following experiments were carried out strictly according to Chinese national standards.

2. Technical requirements of normal-weight sand

NWFA can be distinguished from lightweight fine aggregate (LWFA) by the apparent density (ρ_a) and the bulk density (ρ_b) [1, 2]. If $\rho_a \geq 2500\text{kg/m}^3$ or $\rho_b \geq 1400\text{kg/m}^3$ and the void ratio (e_v) $\leq 44\%$ when calculated by Eq. (1), the sand is NWFA [1]. On the other hand, if $\rho_b \leq 1200\text{kg/m}^3$, the sand is LWFA. Also, the LWFA is usually made of lightweight coarse aggregate (LWCA) for producing lightweight aggregate concrete (LWAC) [2]:

$$e_v = \left(1\text{-}\frac{\rho_b}{\rho_a}\right) \times 100\% \tag{1}$$

In order to standardise the artificial sand (AS), namely, manufactured sand (MS), the terms, definitions, classifications, specifications, technical requirements, test methods, inspection rules, and so on are stipulated in the Chinese national standards Sand for Construction (GB/T14684-2011) [1], Technical Specification for Application of Manufactured Sand Concrete (JGJ/T241–2011) [3], and Standard for Technical Requirements and Test Method of Sand and Crushed Stone (or Gravel) for Ordinary Concrete (JGJ52–2006) [4].

The MS is made of different parent rocks, whose strength should be in accordance with **Table 1** [3].

The fineness module (M_x, calculated by Eq. 2) should be $M_{x,\,c}$ = 3.7–3.1, $M_{x,\,m}$ = 3.0–2.3, $M_{x,\,f}$ = 2.2–1.6, and $M_{x,\,e}$ = 1.5–0.7 for coarse, medium, fine, and extra-fine sand, respectively. The detailed particle size distribution sieved with a square hole mesh sieve is provided in **Table 2**:

$$M_x = \frac{(A_2 + A_3 + A_4 + A_5 + A_6)\text{-}5A_1}{100\text{-}A_1} \tag{2}$$

where A_1, A_2, A_3, A_4, A_5 and A_6 are the cumulative percentages retained in 4.75, 2.36, 1.18, 600, 300 and 150 μm sieves, respectively.

For concrete production, it is recommended that sand from Zone II be used. The sand ratio (sand-to-sand and coarse aggregate weight ratio, S_p, %) should be improved properly when selecting sand from Zone I to keep a sufficient cement content to satisfy the workability requirement of concrete. When selecting sand from Zone III, S_p should be reduced properly. Also, medium sand should be selected

Igneous rock	Metamorphic rock	Sedimentary rock
≥100	≥80	≥60

Notes: The test method of compressive strength of parent rock refers to GB/T50266–2013 [5].

Table 1.
Strength of parent rock made of artificial sand (unit, MPa).

Nominal diameter of sand and sieve hole		5 mm	2.5 mm	1.25 mm	630 μm	315 μm	160 μm
Mesh size (length of sieve hole)		4.75 mm	2.36 mm	1.18 mm	600μm	300μm	150 μm
Cumulative percentage retained (%)	Zone I	10–0	35 − 5	65–35	85–71	95–80	100–90
	Zone II	10–0	25–0	50 − 10	70–41	92–70	100–90
	Zone III	10–0	15–0	25–0	40–16	85–55	100–90

Table 2.
Particle size distribution for artificial sand stipulated in [1, 3].

for self-compacting concrete (SCC) production, and the percentage retained on a 315-μm sieve should not be less than 15%. The fineness modulus range of 2.6–3.0 is suitable for high-strength concrete (HSC) production. Moreover, medium sand should be used for mass concrete and mortar production [6].

3. Technical requirements of lightweight aggregate

Lightweight aggregate (LWA) includes artificial, natural, industrial waste slag, cinder, and spontaneous combustible coal gangue LWA [2]. LWA is called super-lightweight coarse aggregate (SLWCA) when ρ_b does not exceed 500 kg/m^3. The tube crushing strengths (TCS) of the different grades of high-strength lightweight coarse aggregates (HSLWCAs) are provided in **Table 3**. The softening coefficient should be equal to or higher than 0.8 and 0.7 for artificial and industrial waste slag LWCA and natural LWCA, respectively. The fineness modulus ($M_{x,\ LA}$) of LWFA should be between 2.3 and 4.0.

According to the current LWA production technology and its use in actual engineering, the particle size distribution of LWA is shown in **Table 4**.

Bulk density grade (ρ_b, kg/m^3)	**Tube crushing strength (MPa)**	**Strength grade of LWAC**
600	4.0	LC25
700	5.0	LC30
800	6.0	LC35
900	6.5	LC40

Notes: The bulk density grade is a size range, not an exact number. For example, ρ_b = 600 kg/m^3 means 500 < ρ_b 600 kg/m^3 and so on.

Table 3.
Tube crushing strength and strength grade of LWAC for artificial HSLWCA stipulated in [2].

Mesh size (length of sieve hole)	4.75 mm	2.36 mm	1.18 mm	600μm	300μm	150 μm
Cumulative percentage retained (%)	0–10	0–35	20–60	30–80	65–90	75–100
Continuous grading (5–16 mm)	16.0 mm	9.50 mm	4.75 mm	2.36 mm		
Cumulative retained percentage (%)	0–10	20–60	90–100	95–100		
Continuous grading (5–10mm)	9.50 mm	4.75 mm	2.36 mm			
Cumulative percentage retained (%)	0 − 15	90–100	95–100			

Table 4.
Particle size distribution for LWFA and LWCA.

4. Lightweight aggregate concrete

All-lightweight aggregate concrete (ALWAC), also known as all-lightweight concrete (ALWC), is made from LWFA, LWCA, cement, water, and other admixtures. Although ALWC has many excellent properties, such as high specific strength (ratio of cubic compressive strength to dry apparent density, *SS*, kN·m/kg), high anti-deformability, excellent fire resistance, and so on, there are a number of negative aspects, such as low elastic modulus, the fact that LWCA rises more easily when pumping and vibrating LWAC, high costs, and so on. In order to satisfy the quality of both normal-weight concrete (NWC) and ALWC, new types of LWACs are formed by replacing parts of LWA with normal-weight aggregate (NWA) and vice versa. The naming rules are as follows.

The term 'ALWAC' was first mentioned in 1972 based on the available literature in the EI and SCI databases [7]. If, on the basis of ALWAC, only a part of LWFA is replaced with normal-weight sand in the same volume ratio S_S (%, $0 < S_S \leq 100\%$), the LWAC is called sand lightweight concrete (SLWC) [7]. Similarly, if only a part of LWCA is replaced ($0 < S_G \leq 100\%$) with normal-weight gravel, the concrete is called gravel lightweight concrete (GLWC). When both lightweight fine and coarse aggregates are replaced at the same time, the concrete is named hybrid aggregate lightweight concrete (HALWC). On the other hand, if on the basis of NWC, only a part of normal-weight coarse aggregate is replaced with expanded ceramsite ($0 < S_C < 100\%$), the concrete is named hybrid aggregate concrete with less ceramsite (HACC). The other corresponding SCCs, fibre-reinforced concrete (FRC) and reinforced concrete (RC), obey the rules too.

Compared to ALWC and NWC, the abovementioned concrete can be uniformly named semi-lightweight concrete (semi-LWC), where the term 'semi' means the LWA is less than half, just half, or more than half in volume. But according to [8], the concrete is called LWAC when the dry apparent density of concrete (ρ_d, kg/m^3) is less than or equal to 1950 kg/m^3. Otherwise, it should be called specified density concrete (SDC) when $1950 < \rho_d \leq 2300$ kg/m^3. The grade of dry apparent density is ranked from 600 to 1900 kg/m^3, the gradation is 100 kg/m^3, and the density range is ± 50 kg/m^3. For instance, ρ_d is 600 kg/m^3 and the range is 560–650 kg/m^3.

In this study, the LWFA and LWCA are shale pottery (SP) and shale ceramsite (SC), while the NWFA and NWCA are MS or natural sand (NS, also known as river sand) and crushed stone (CS), respectively. Both LWCA and NWCA are crushed aggregates. The technical parameters of LWA and normal-weight aggregate (NWA) are listed in **Tables 5** and **6**, respectively.

The porosity of SC and SP is 51.3 and 23.9%, and the void ratio is 41.7 and 47.0% according to tests following GB/T17431.2-2010 [9], respectively. So the SC must be pre-wetted 24 hours (h) before production of concrete in order to prevent reabsorption of mixing water, because the water absorption rate (%, ω_a) is stable after 24 h, as shown in **Table 6**.

The ALWC is more sensitive to mix design than NWC mainly because of the higher porosity, lower bulk density, and lower tube crushing strength (TCS). Although the interface between crushed angular LWA and mortar is very good, the TCS is significantly lower compared to the mortar, and the LWCA floats more easily. The internal structure of ALWC becomes non-uniform, and thus the strength of ALWC depends on the strength of the mortar.

Many experiments have shown that the strength of ALWC mainly depends on the mass of the maximum diameter of LWCA, the mass of cementitious material (cement, fly ash (FA), silica fume, and other admixtures with gelling capacity), the water-to-binder weight ratio (*W*/*B*), and the fine aggregate to overall aggregate weight ratio (%, S_p) (42–47% in general).

SC	>16 mm (%)	16.0 mm (%)	9.50 mm (%)	4.75 mm (%)		
GB/T17431.1–2010	≤ 5	≤ 10	20–60	85–100		
Experimental values	0.1	1.6	34.2	99.7		
	Tube crushing strength (TCS) (MPa)			Mean	Over mean	GB/ T17431.1–2010
≥9.50 mm	3.63	3.67	3.68	3.66	3.62	2.0–3.0
≥4.75 mm	3.54	3.58	3.59	3.57		
SP	4.75 mm (%)	2.36 mm (%)	1.18 mm (%)	0.6 mm (%)	0.3 mm (%)	≤0.15 mm (%)
GB/T17431.1–2010	≤10	≤35	20–60	30–80	65–90	75–100
Experimental values	2.5	11.6	39.8	58.9	69.1	99.8
Gravel	>16 mm (%)	16.0 mm (%)	9.50 mm (%)	4.75 mm (%)	2.36 mm (%)	
GB/T 14685–2011	0	0–10	30–60	85–100	95–100	
Experimental values	0	8.4	48.6	92.3	98.7	
MS	4.75 mm (%)	2.36 mm (%)	1.18 mm (%)	0.6 mm (%)	0.3 mm (%)	≤0.15 mm (%)
GB/T 14684–2011	0–10	0–25	10– 50	41–70	70–92	80–94
Experimental values	7.8	24.6	47.2	66.9	89	93.7
NS	4.75 mm (%)	2.36 mm (%)	1.18 mm (%)	0.6 mm (%)	0.3 mm (%)	≤0.15 mm (%)
GB/T 14684–2011	0 − 10	0–25	10–50	41–70	70–92	90–100
Experimental values	6.5	21.4	37.9	63.9	89.9	97.9

Notes: The fineness module values of MS and NS are 3.06 and 2.98, respectively.

Table 5.
Technical parameters of LWA and NWA stipulated in the Chinese national standard and test values.

Diameter range (mm)	**0.5 h**	**1 h**	**2 h**	**4 h**	**6 h**	**8 h**	**12 h**	**24 h**	**32 h**	**48 h**	**72 h**
5–8	7.5	8.6	8.8	8.9	9.3	9.5	9.8	10.4	10.5	10.6	10.6
8–15	6.6	7.3	7.5	7.6	7.7	8.0	8.1	8.7	8.8	8.9	8.9

Table 6.
Water absorption rate (ω_a) of SC after different numbers of hours (h).

Generally, the strength of NWC increases with curing time; however, the strength of ALWC decreases when the diameter of LWCA is larger than 20 mm. Because of that, the maximum diameter of LWCA should be 15 mm or smaller. On the other hand, the SC has a softening effect, the softening coefficient (Ψ_s) stipulated in GB/T17431.1–2010 [2] is not less than 0.8, and increasing the maximum diameter leads to a decrease in Ψ_s. The test result is shown in **Table 7**, which can explain the difference in the strength forming mechanism between ALWC and NWC. Besides the abovementioned, when the maximum diameter of LWCA is larger, the damage area (area of LWCA versus mortar in a cross-section) is larger, so the strength of ALWC is lower; for example, the cubic compressive strength of ALWC is 32 MPa in 28 days, while the mortar strength after removing all LWCA in fresh concrete is 45 MPa.

Diameter range (mm)	7 d	14 d	28 d	60 d	90 d	120 d	180 d
Ψ_{s1} (saturated surface dry condition)							
5–8	0.86	0.81	0.77	0.72	0.66	0.58	0.55
8–15	0.92	0.90	0.86	0.81	0.75	0.72	0.70
Ψ_{s2} (oven dry condition)							
5–8	0.88	0.85	0.83	0.81	0.79	0.78	0.77
8–15	0.96	0.94	0.92	0.90	0.89	0.88	0.87

Table 7.
Softening coefficient (Ψ_s) of SC after soaking in water for different numbers of days (d).

Type of concrete	m_C (kg)	m_{FA} (kg)	m_{SC} (kg)	m_{CS} (kg)	m_{SP} (kg)	m_{MS} (kg)	m_W (kg)	f_{cu} (MPa)	f_c (MPa)	f_{ts} (MPa)	E_c (GPa)	ρ_d (kg/m^3)
ALWC	481	157	444	—	408	—	171	29.3	28.6	2.32	14.56	1594
SLWC			444	—	367	70		30.5	29.7	2.47	16.54	1612
GLWC			311	339	408	—		32.2	32.0	3.01	17.26	1796
HALWC			333	283	367	70		31.8	31.3	2.81	16.86	1785
HACC*	300	128	45	957	—	624	235	37.3	24.6	3.90	36.47	2280

Notes: (1) m_C, m_{FA}, m_{SC}, m_{CS}, m_{SP}, m_{MS}, and m_W stand for cement (C), fly ash (FA), shale ceramsite (SC), crushed stone (CS), shale pottery (SP), manufactured sand (MS), and water (W), respectively; (2) f_{cu}, f_c, and f_{ts} stand for the cubic compressive strength, axial compressive strength (prism specimen, height width ratio is 2 or 3), and splitting tensile strength (cubic specimen) at 28 days, respectively; (3) E_c is Young's elastic modulus; (4) ρ_d stands for dry apparent density; (5) HACC is the specified density concrete (SDC) judged by ρ_d.
**HACC is the specified density concrete (SDC), the symbol of concrete strength grade can be expressed by "SC", which can be different from the symbol "C" of normal-weight concrete (NWC), and "LC" of lightweight aggregate concrete (LWAC).*

Table 8.
Reference mixes (1 m^3) and test results of LWACs and SDC for LC30.

All of the following concretes are designed by pumping concrete; that is, the slump is 160–220 mm, mainly taking LC30, for example. The reference mixes and test results are shown in **Table 8**.

In **Table 8**, the cementitious material is PO42.5 Portland cement and Grade II fly ash. The water reducing rate of high performance water reducing agent is not less than 20% and added 1.6–2.0 wt% (by mass of cementitious material).

Because of the difference of LWA and NWA, the strength, elastic modulus, and dry apparent density are increased when LWFA or LWCA is replaced separately by normal-weight aggregates, but it is more complex when replaced at the same time.

Also, the ratio (ζ) of axial compressive strength to cubic compressive strength for LWAC is usually close to 1.0, which is larger than the value of 0.66–0.67 required by JGJ51-2002 [6] and also larger than the value of 0.76 given for NWC. This phenomenon is precisely because the LWA with lower TCS will be crushed before the cement mortar and will show larger deformation, which is equivalent to antifriction and thus with self-lubricated capability.

4.1 Autogenous shrinkage properties of LWACs

Because the hydration reaction of cement is an exothermic process, the amount of heat released leads to a temperature difference both inside and outside the concrete, and the temperature stress induces the appearance of cracks.

	ALWC	SLWC	GLWC	HALWC
Time (h)	31.0	34.0	36.0	27.5
Maximum temperature (°C)	66.4	63.9	62.4	65.5
Calculated temperature (°C)	66.1	69.7	59.5	53.5

Table 9.
Test and calculated values of adiabatic temperature rise for LWACs.

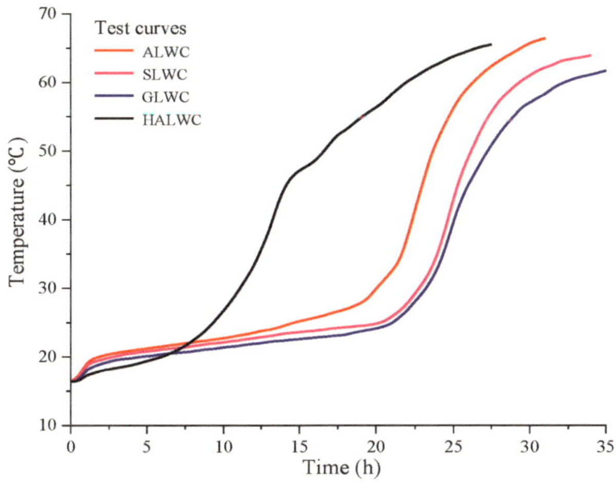

Figure 1.
Relationships between adiabatic temperature rise and time.

According to GB50496-2009 [6], the adiabatic temperature rise of LWACs is shown in **Table 9** (tested with a 3.7-litre Thermos bottle) and **Figure 1**.

Because the gas pressure in the Thermos bottle becomes higher as the hydration reaction proceeds, the Thermos glass liners burst at a certain time, as shown in **Table 9**. However, the length of time is shortest for HALWC. The reason may be the uniform distribution of normal-weight fine and coarse aggregates, which can improve the heat conduction rate and provide a better temperature distribution. On the contrary, the time period of the temperature rise is shorter for SC and SP compared to SLWC and GLWC, because of the higher porosity of LWA, which provides a better insulation performance. The difference between SLWC and GLWC may be that the distribution of NWFAs is more uniform than that of coarse aggregate, such as the sand particles touched in dot form can speed up heat conduction.

Autogenous shrinkage of concrete mainly happens after the initial setting time, but chemical shrinkage, which includes three complete stages, also has a significant influence. According to the mechanism of concrete shrinkage, chemical shrinkage happens because the absolute volume of hydration products is smaller than that of water and binding material during the early stage. Autogenous shrinkage happens during the skeleton structure forming in a later stage, so the unhydrated cement particles react further. In fresh concrete, the volume can also cause shrinkage because of the setting of parts of particles. Drying shrinkage is caused by water loss. In short, the cracks in concrete are mainly caused by plastic shrinkage in the early stage.

In **Figure 2**, the curves are smoothed after the peak values because of higher fluctuation (shown by dashed lines). The higher porosity and rougher surfaces of

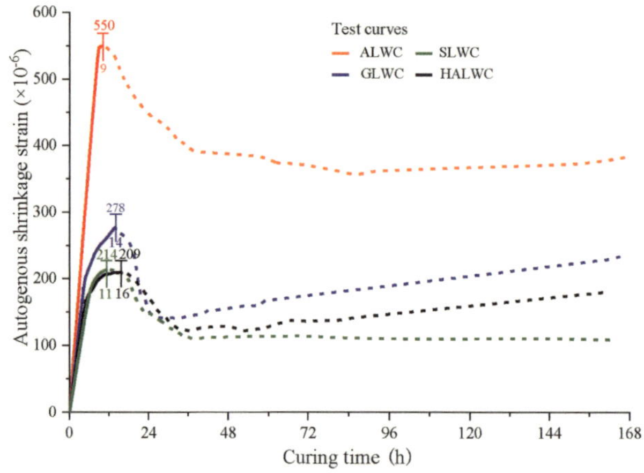

Figure 2.
Test curves of autogenous shrinkage strain with time.

SC and SP also result in larger specific surface areas (total area of material per unit mass, m^2/g), which can absorb more cement particles and thus improve hydration conduction, so the autogenous shrinkage of ALWC is greater in a shorter time period and then becomes stable. Among SLWC, GLWC, and HALWC, the autogenous shrinkage is mainly determined by the amounts of LWA. Because the specific surface area of aggregates is different, the internal distribution of aggregates is uniform, and parts of cement particles are subsident, which can determine the internal temperature stress field, so the resistance capability of plastic deformation is different.

4.2 Durability properties of LWACs

The effects of different mineral admixtures on the durability of LWACs were studied. **Table 10** shows the effects of substituting 75 wt% mineral powder (denoted as MP75; the activity index is 96% in 28 days) for fly ash and 50 wt% limestone powder (denoted as LP50), respectively. And both mineral powder and limestone powder are mixed in a ratio of 1:1, and total substitution for fly ash (marked MP75 + LP50) is based on **Table 8** (with a slight difference). The test method is according to GB50082–2009 [10].

Mineral powder can enhance both the strength and durability of concrete. Although the added limestone powder only reduces the strength of concrete, the requirements can be met, and the cost can be reduced. Because fly ash has become a scarce resource in China, mineral and limestone powder can be an effective alternative in the ready-mixed concrete industry.

Generally, the effects of normal-weight aggregates and mineral admixtures on carbonation and electric flux are not obvious, but according to GB50082-2009 [10], when the electric flux is between 1000 and 2000 C, the grade of chloride iron penetration is low, so the concretes can meet the requirements of the code.

4.3 Softening properties of LWACs

Since SC has a softening effect according to **Table 7**, is there also an SC concrete softening effect? The test results (**Table 11**) show that the LWACs almost have no softening effects. A possible reason is that the LWAs are strengthened because of their absorption of cement particles and hydration; on the other hand, the main contribution to the strength comes from the cement mortar, which does not show softening.

Type	$h_c^{28\,d}$ (mm)	$Q_e^{28\,d}$ (C)	Δm^{21} (g)	f_{cu}^{21} (MPa)	$f_{cu}^{28\,d}$ (MPa)	K_f (%)
ALWC	12.6	1235	31.3	30.1	31.2	96.5
ALWC-MP_{75}	10.8	1126	28.6	32.6	33.4	97.6
ALWC-LP_{50}	11.8	1226	46.8	28.2	32.6	86.5
ALWC-MP_{50} + LP_{50}	—	—	33.7	29.8	34.7	85.9
SLWC	11.2	1126	32.6	36.6	38.6	94.8
SLWC-MP_{75}	9.3	1042	31.6	38.1	40.6	93.8
SLWC-LP_{50}	10.4	1092	43.1	34.3	39.4	87.1
SLWC-MP_{50} + LP_{50}	—	—	—	—	40.1	—
GLWC	10.2	1326	35.5	41.2	44.5	92.6
GLWC-MP_{75}	8.3	1265	34.2	43.7	45.4	96.3
GLWC-LP_{50}	9.5	1301	44.5	38.6	46.5	83.0
GLWC-MP_{50} + LP_{50}	—	—	—	—	46.1	—
HALWC	12.9	1410	34.7	36.9	39.7	92.9
HALWC-MP_{75}	10.6	1339	33.5	38.3	40.8	93.9
HALWC-LP_{50}	11.2	1389	46.3	35.2	40.5	86.9
HALWC-MP_{50} + LP_{50}	—	—	—	—	40.8	—
HACC	9.0	1007	—	36.4	38.3	95.0

Notes: (1) $h_c^{28\,d}$ stands for the carbonation depth of concrete at 28 d; (2) $Q_e^{28\,d}$ stands for electric flux after 6 h under chloride ion penetration of concrete at 28 d; (3) Δm^{21} and f_{cu}^{21} stand for the mass loss and cubic compressive strength of concrete after 21 dry-wet cycles under sulphate attack, respectively; (4) the test is stopped when the ratio of $K_f = f_{cu}^{21}/f_{cu}^{28\,d} \times 100\%$ is larger than 75% according to GB50082–2009 [10].

Table 10.
Test results of durability for different LWACs and SDC with LC_{30}.

Table 11 shows that the concrete strength and elastic modulus increase under curing in water on different days. However, the HALWC falls slightly, and the elastic modulus is almost constant. Similarly, for any type of LWAC, all of the ζ values are also close to 1.0 on different days.

The test results show the LWACs are without a softening effect, so they can be used in hydraulic structure engineering and underground engineering.

4.4 Properties of LWACs after elevated temperature treatment

The appearance characteristics and strength of LWACs are shown in **Tables 12** and **13** under different temperatures (T, °C), respectively. When the temperature is below 200°C, the surface of concrete shows no change, and the strength increases, but when the temperature is above 300°C, changes in both colour and crack shape can be observed. The maximum width of cracks (w_{max}, mm) increases with increases in temperature and the strength decreases.

Although the mass loss in different types of concretes shows no obvious difference after high-temperature treatment, the effect of the addition of NWAs alone on the strength and elastic modulus is higher and changes regularly; that is, added NWCA alone larger than added NWFA alone, but smaller when added at same time than added NWCA only.

In general, the residual values of elastic modulus are larger than those of strength, which means the anti-deformation capacity of concrete decreases with

	Type	28 d	60 d	90 d	120 d	180 d	ψ_c (%)
f_{cu} (MPa)	ALWC	30.98	31.54	31.05	30.15	29.78	101.3
	SLWC	31.48	32.43	31.87	31.44	31.20	100.9
	GLWC	34.43	35.18	35.09	34.92	34.81	102.8
	HALWC	33.56	34.22	33.96	33.68	33.47	102.5
f_c (MPa)	ALWC	29.45	30.27	29.81	28.95	28.07	99.9
	SLWC	30.87	31.56	30.78	30.49	30.15	100.6
	GLWC	33.59	34.14	33.96	33.75	33.62	102.7
	HALWC	31.35	32.84	32.61	32.39	32.18	102.9
f_{ts} (MPa)	ALWC	2.96	3.03	2.95	2.86	2.79	104.5
	SLWC	3.04	3.11	3.03	2.95	2.90	104.3
	GLWC	3.15	3.23	3.20	3.18	3.15	104.3
	HALWC	3.07	3.19	3.14	3.11	3.09	105.1
E_c (GPa)	ALWC	21.23	21.74	22.35	23.24	23.79	115.2
	SLWC	21.62	22.24	22.85	23.41	23.98	115.3
	GLWC	21.92	22.80	23.35	23.96	24.71	116.4
	HALWC	21.58	22.45	23.09	23.62	24.35	116.3

Notes: ψ_c stands for the softening coefficient of concrete.

Table 11.
Test results of LWACs with LC30 under water curing on different days.

T (°C)	Colour	Visible phenomenon	w_{max} (mm)			
			ALWC	SLWC	GLWC	HALWC
300	Light grey	Fewer hairline cracks	0.06	0.04	0.05	0.04
400	Off-white	More hairline cracks	0.16	0.12	0.14	0.12
500	Hazel	Honeycomb cracks	0.20	0.20	0.20	0.18
600	Brownness	Honeycomb cracks with surface wrapping	0.30	0.26	0.28	0.24

Table 12.
Appearance characteristics of LWACs after high-temperature treatment.

rises in temperature. On the other hand, the residual strengths of LWCAs after high-temperature treatment are very close to or even higher than that of NWC, which indicates that LWACs can be used for fire-resistant design. For example, the residual strength of axial compression is 95% at 200°C, 80–90% at 300°C, 70–75% at 400°C, 60–65% at 500°C, and around 50% at 600°C. The residual splitting tensile strength is around 90%, 75–80%, 60–65%, 50–55%, and around 45% at 200, 300, 400, 500, and 600°C, respectively.

4.5 Properties of LWACs cured at negative temperature

During the construction process used by the artificial freezing method, the ambient temperature in the working place is from −8 to −12°C in China. Concrete properties after curing at negative temperature are shown in **Tables 14** and **15**, where the specimen is wrapped with a layer of quilt after being poured and then put into a low-temperature test chamber.

	Type	200°C	300°C	400°C	500°C	600 °C	η_c^T (%)
Δm (%)	ALWC	1.1	4.5	5.4	6.4	7.0	
	SLWC	1.3	4.6	5.6	6.7	7.2	
	GLWC	0.9	4.2	4.7	5.5	6.8	
	HALWC	0.8	4.2	4.9	5.8	6.9	
f_{cu} (MPa)	ALWC	30.54	25.70	22.02	19.79	16.48	55.9
	SLWC	32.05	28.98	25.01	21.35	17.22	56.4
	GLWC	33.23	32.25	25.86	23.05	18.34	57.0
	HALWC	32.65	31.82	25.14	22.21	17.90	56.8
f_c (MPa)	ALWC	26.88	23.43	20.04	17.50	14.68	52.0
	SLWC	26.76	25.31	21.82	18.90	15.12	50.9
	GLWC	29.77	26.96	23.19	19.14	15.95	50.9
	HALWC	28.51	26.92	22.65	18.51	14.82	48.1
f_{ts} (MPa)	ALWC	2.06	1.79	1.46	1.25	1.07	46.1
	SLWC	2.22	1.93	1.59	1.36	1.14	46.2
	GLWC	2.73	2.40	1.96	1.69	1.47	48.8
	HALWC	2.53	2.25	1.80	1.55	1.35	48.0
E_c (GPa)	ALWC	12.61	10.13	8.19	6.62	5.03	34.5
	SLWC	15.68	13.75	10.98	8.88	6.13	37.1
	GLWC	16.89	16.46	14.53	11.19	8.76	50.8
	HALWC	16.11	15.32	13.65	10.11	8.12	48.2

Notes: Δm stands for the ratio of mass after high-temperature treatment to that under room temperature. η_c^T stands for the ratio of strength or elastic modulus at 600°C to that at room temperature.

Table 13.
Test results of LWACs for LC30 after elevated temperature treatment.

The compressive strength and elastic modulus of the specimens cured at −5°C meet the basic requirements (up to 90%) both with and without anti-freezing agent, but at −10°C, they can satisfy the requirements when added anti-freezing agent.

Although the strength and elastic modulus of other groups fail the basic requirements, the hydration reaction does not stop but only diminishes. Compared with NWC, the strength of the specimen is higher, which shows that it helps promote the hydration reaction because of the heat preservation and inner curing effect of LWA. On the other hand, the strength of GLWC is slightly higher than that of ALWC under the same conditions. The reason is the same as the one mentioned above, namely, that the elastic modulus of gravel is larger than that of ceramsite.

Also, fibre can enhance the strength and elastic modulus of specimens cured at negative temperature, and the laws are also the same; that is, the performance of concrete is better with elastic modulus of fibre increasing.

To summarise, the LWACs can meet the requirements of freezing process construction.

4.6 Uniaxial stress-strain curves of LWACs

The uniaxial stress-strain curve of LWACs is similar to that of NWC, as shown in **Figure 3**, but the total strain of LWACs is significantly larger than that of NWC.

T (°C)	w_A (%)	ALWC, f_{cu} (MPa)						η_c^T (%)	GLWC, f_{cu} (MPa)				η_c^T (%)
		1 d	2 d	3 d	7 d	14 d	28 d		3 d	7 d	14 d	28 d	
–5	0	4.9	11.3	16.5	20.5	24.7	29.3	90.1	16.9	21.1	26.0	31.3	90.9
	2	6.2	12.5	17.9	27.3	—	32.1	98.7	—	—	—	—	—
−10	0	4.1	10.8	13.8	17.5	21.4	25.0	76.9	14.4	18.9	23.3	27.2	79.0
	2	5.8	12.1	16.1	24.4	—	29.7	91.3	—	—	—	—	—
−15	0	37	8.5	10.3	13.3	15.8	18.6	57.2	11.4	14.3	17.5	20.7	60.1
	3	5.3	11.8	15.4	18.9	—	27.5	84.6	—	—	—	—	—

*Notes: (1) w_A stands for the ratio of anti-freezing agent to cement (by mass); (2) η_c^T stands for the ratio of strength at 28 days and curing at negative temperature to that at room temperature; (3) the mixes are slightly different from **Table 8**.*

Table 14.
Compressive strength of LWACs cured at negative temperature with anti-freezing agent for LC30.

T (°C)	ALWC, f_c (MPa)				η_c^T (%)	GLWC, f_c (MPa)				η_c^T (%)
	3 d	7d	14 d	28 d		3 d	7d	14 d	28 d	
–5	14.46	18.41	22.36	24.63	83.0	15.81	20.34	24.64	26.45	85.0
−10	12.65	16.16	19.63	21.34	71.9	14.86	17.87	21.98	23.34	75.0
−15	10.88	13.18	15.54	16.62	56.0	11.94	14.49	17.21	18.15	58.3
	f_{ts} (MPa)					f_{ts} (MPa)				
-5	1.87	2.49	3.03	3.42	91.9	2.03	2.54	3.13	3.49	92.3
–10	1.69	2.15	2.71	2.99	80.4	1.83	2.30	2.75	3.06	81.0
–15	1.50	1.88	2.21	2.46	66.1	1.69	1.97	2.32	2.55	67.5
	E_c (GPa)					E_c (GPa)				
-5	9.43	13.22	16.49	18.77	92.0	10.02	14.23	17.56	19.36	94.0
–10	8.81	12.58	15.03	16.52	81.0	8.97	12.01	14.98	17.10	83.0
–15	8.26	11.57	13.61	14.28	70.0	8.44	11.15	13.22	15.04	73.0

Table 15.
Test results of LWACs cured at negative temperature without anti-freezing agent for LC30.

The symbols of stress and strain obey the following rules: the plus sign '+' denotes tension; the minus sign '–' denotes compression.

Generally, the stress-strain curve can be expressed by Eqs. (3) and (4) [11].

Ascending curve:

$$y = \alpha x + (3\text{-}2\alpha)x^2 + (\alpha\text{-}2)x^3 \quad x \leq 1 \tag{3}$$

Descending curve:

$$y = \frac{x}{\beta(x-1)^2 + x} \quad x \geq 1 \tag{4}$$

where $x = \frac{\varepsilon}{\varepsilon_0}$ and $y = \frac{\sigma}{\sigma_0}$. α and β are fitting coefficients shown in **Table 16.** ε_0 and σ_0 stand for peak strain and peak stress under uniaxial compression, respectively.

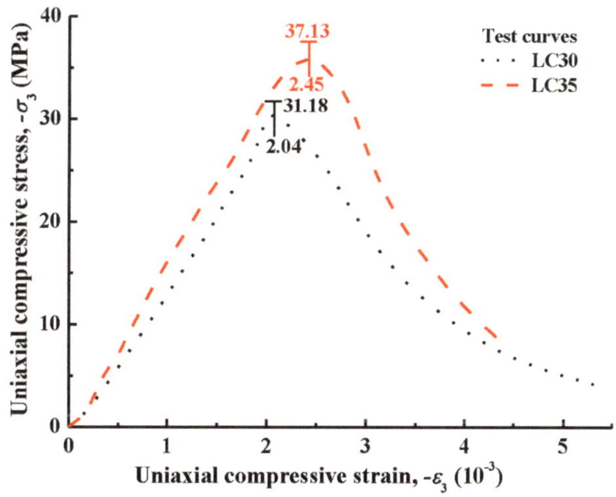

Figure 3.
Test curves of stress-strain under uniaxial compression for ALWC.

Strength grade	Ascending curve		Descending curve	
	α	R^2	β	R^2
LC 30	−0.3777	0.9332	4.7704	0.9921
LC 35	0.4398	0.9855	9.0217	0.9975

Table 16.
Fitting coefficients and relative coefficients in Eqs. (3) and (4).

Eqs. (3) and (4) can be fitted for all kinds of concretes, whether or not the curve is complete. In particular, direct measurement of the descending curve is not usually easy (**Figure 4**).

Although **Figure 4** does not contain descending curves, the law of different LWACs is the same; that is, the ultimate stress decreases, and the ultimate strain increases with the temperature increase, which shows that the plastic deformation gets larger because the strength of the cement mortar matrix decreases. On the other hand, under the same temperature, the effects of a small quantity of NWA on the ultimate strain are not significant; only the effect on the ultimate stress is remarkable. At the same time, ALWC is similar to SLWC, and GLWC is similar to HALWC.

4.7 Multiaxial strength of LWACs

Under multiaxial compressive stresses, the ultimate compressive strength of concrete will increase significantly, and therefore the failure modes change. For example, ALWC undergoes the phenomenon of squeeze flow, and a plastic plateau appears in the stress-strain curve under the two larger lateral stresses.
Tables 17 and **18** show the bi- and triaxial ultimate compressive strengths tested by a large real triaxial test system, respectively. To reduce the friction, a two-layer polythene film with lithium base oil smeared between the layers is used, which can guarantee that the strength under single stress action (σ_0) is close to the axial compressive strength (f_c). In the test, σ_0 is slightly smaller than f_c, and the loading type is proportional loading. The samples were tested after 120 days of curing.

The formulas for calculating bi- and triaxial ultimate strength are shown as Eqs. (5) and (6) [12], respectively:

$$\sigma_3 = \frac{1+\omega_2}{1+\omega_2^{-1}}\sigma_2, \omega_2 = \frac{\sigma_3}{\sigma_2} \tag{5}$$

$$\sigma_3 = \frac{\sqrt{\sigma_1\sigma_2(1+\omega_1+\omega_3)\left(1+\omega_2+\omega_3^{-1}\right)}}{1+\omega_1^{-1}+\omega_2^{-1}} \quad \omega_1 = \frac{\sigma_3}{\sigma_1}, \omega_3 = \frac{\sigma_2}{\sigma_1} \tag{6}$$

The smaller relative error indicates that the test data are reliable and the formulas are correct. On the other hand, the multiaxial ultimate strength increases with

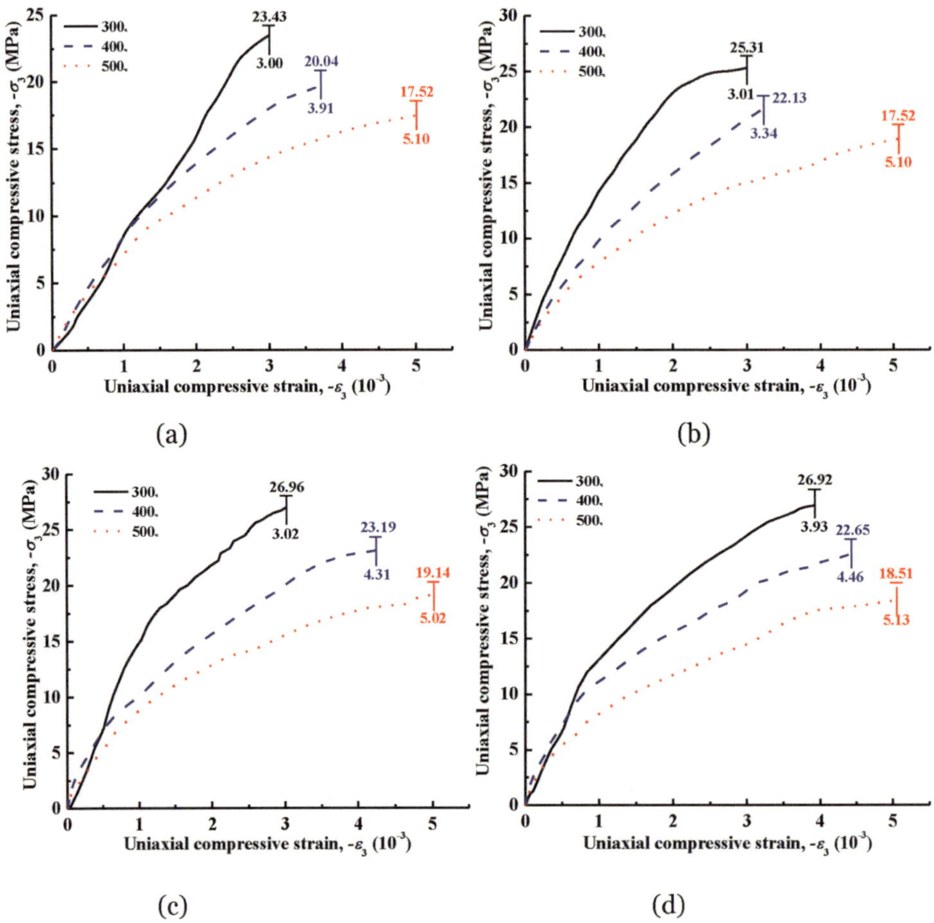

Figure 4.
Test curves of stress-strain under uniaxial compression for LWACs after elevated temperature treatment. (a) ALWC, (b) SLWC, (c) GLWC, and (d) HALWC.

σ_3:σ_2	$-\sigma_{30}$ (MPa)	Eq. (5) (MPa)	E_r (%)	$\sigma_{30}/f_c^{120\ d}$
1:0.25	40.94	−42.62	4.10	1.28
1:0.5	39.92	−41.75	4.58	1.24
1:0.75	39.88	−41.96	5.22	1.25
1:1	41.88	−44.28	5.73	1.31

Notes: (1) σ_{30} stands for peak stress, namely, ultimate compressive strength. E_r stands for relative error.

Table 17.
Biaxial compressive strength under proportional loading.

σ_3:$-\sigma_2$:$-\sigma_1$	$-\sigma_1$ (MPa)	$-\sigma_2$ (MPa)	$-\sigma_{30}$ (MPa)	Eq. (6) (MPa)	E_r (%)	$\sigma_{30}/f_c^{120\ d}$
1:0.1:0.1	5.11	4.98	49.77	−51.1	2.6	1.56
1:0.25:0.25	19.58	19.10	75.85	−78.32	3.2	2.37
1:0.5:0.5	50.61	49.21	97.94	−101.22	3.3	3.06
1:1:0.1	6.48	60.81	60.51	−64	5.7	1.89
1:1:0.25	22.56	87.03	86.48	−90.24	4.3	2.70
1:1:0.5	55.61	107.37	106.78	−111.22	4.1	3.34
1:0.25:0.1	7.56	36.62	58.62	−60.60	3.3	1.83
1:0.5:0.1	19.04	30.98	72.85	−75.80	4.1	2.28
1:0.5:0.25	6.06	14.77	73.64	−76.16	3.4	2.30

Table 18.
Triaxial compressive strength under proportional loading.

increasing lateral stress; the law is the same as for NWC, but the ratio of σ_{30} to f_c is larger than that of NWC. Also, the stress-strain curves show that the deformation resistance capacity of LWACs is stronger than that of NWC, so the LWACs cannot be crushed easily and thus have higher strength under the action of multiaxial stress.

The ultimate strength and elastic modulus of LWACs under traditional triaxial stresses are shown in **Table 19**.

All of the strength and elastic modulus values increase with increasing confining pressure. Under the same temperature and confining pressure, the effect on NWA is highest when using the lowest amount of hybrid aggregate, such as gravel, NS, and LWS. However, when the temperature is above 300°C, the strength of HALWC is smaller than those of SLWC and GLWC. At the same time, below 300 °C, the strength increases, except for HALWC, whose strength increases at temperatures below 200°C. The relationship between ultimate compressive strength and confining pressure can be expressed by Mohr-Coulomb theory as shown in Eq. (7).

	T (°C)	−2 MPa	−4 MPa	−6 MPa	−8 MPa	−10 MPa
ALWC $-\sigma_{30}$ (MPa)	20	29.30	32.92	35.25	39.29	43.23
	200	31.04	37.45	42.98	48.10	51.86
	300	26.96	31.57	37.55	40.89	44.76
	400	22.15	27.30	32.77	38.14	40.83
	500	25.90	27.04	30.58	32.57	36.26
SLWC $-\sigma_{30}$ (MPa)	20	31.06	34.23	37.97	39.49	44.50
	200	36.36	40.56	43.54	45.90	50.27
	300	40.76	42.41	49.27	51.07	53.70
	400	30.80	33.48	39.63	43.33	48.22
	500	30.66	32.73	37.52	42.04	43.74
GLWC $-\sigma_{30}$ (MPa)	20	37.44	44.89	49.84	55.19	59.23
	200	40.59	47.56	51.44	57.85	62.10
	300	41.08	47.51	54.31	60.44	63.99
	400	34.74	40.07	45.90	46.66	54.76
	500	30.91	35.71	38.59	45.81	49.18

	T (°C)	−2 MPa	−4 MPa	−6 MPa	−8 MPa	−10 MPa
HALWC $-\sigma_{30}$ (MPa)	20	34.73	40.61	45.47	52.27	53.87
	200	43.34	45.55	49.35	55.42	59.59
	300	42.51	43.42	46.14	50.83	53.79
	400	30.84	37.71	42.13	47.56	51.81
	500	26.55	31.24	37.63	43.07	49.63
ALWC E_c (GPa)	20	10.74	11.03	11.21	12.86	14.15
	200	10.94	11.62	13.31	13.82	14.36
	300	6.29	7.64	7.51	8.19	11.12
	400	5.25	6.64	6.86	7.65	7.78
	500	4.97	5.37	6.13	6.23	6.37
SLWC E_c (GPa)	20	9.80	10.43	10.56	12.42	14.52
	200	12.21	12.83	13.22	13.70	14.83
	300	10.72	10.89	12.26	12.75	14.35
	400	7.27	8.53	8.67	8.72	8.80
	500	6.80	6.93	7.69	7.89	8.21
GLWC E_c (GPa)	20	10.86	11.56	13.23	14.62	15.75
	200	9.10	13.36	13.76	14.35	14.47
	300	10.95	11.12	13.66	13.75	14.01
	400	8.55	8.86	8.90	9.83	9.97
	500	7.01	7.23	7.54	7.45	8.15
HALWC E_c (GPa)	20	11.18	13.70	14.12	14.78	15.90
	200	11.80	12.33	12.55	14.01	15.09
	300	10.98	12.01	12.31	13.96	14.19
	400	7.78	8.84	8.93	9.81	9.82
	500	6.69	7.27	8.13	8.24	8.84

Notes: The values of confining pressure are 2, 4, 6, 8, and 10 MPa, respectively.

Table 19.
Test results of LWACs under traditional triaxial compression after elevated temperature treatment for LC30.

$$\frac{\sigma_{30}}{f_c} = 1 + C\left(\frac{\sigma_1}{f_c}\right)^c \quad \sigma_1 = \sigma_2 \tag{7}$$

where C and c are the fitting coefficients for each temperature group and concrete group, respectively.

The absolute values of relative error are all smaller than 5%.

5. Properties of lightweight sand foamed concrete

In China, traditional foamed concrete generally consists of cement, NS, water, foam agent, and so on [13, 14]. Because the densities of cement and NS are significantly higher than the density of water, these particles sink easily and therefore

Type	m_C (kg)	m_{FA} (kg)	m_{SP} (kg)	m_{SC} (kg)	m_W (kg)	m_F (g)	f_{cu}^{28d} (MPa)	ρ_d (kg/m^3)	λ (w/(m.k))
LFC 5	230	130	430	110	112	407	5.2	973	0.12
LFC 10	260	145	520	130	126	370	12.1	1185	0.26
LFC 20	360	198	543	136	173	280	24.5	1410	0.35
LFC 30	360	235	630	220	184	208	33.1	1723	0.42

Notes: m_F stands for the mass of foam agent.

Table 20.
Reference mixes (1 m^3) and test results of all-lightweight foamed concrete.

cause the foamed concrete to crack. According to the properties, the bulk density of LWA is smaller than that of water, and thus the foamed concrete consists of LWA foamed by physical foaming, which can be called all-lightweight foamed concrete (ALWFC). It does not crack and also makes a higher-strength-grade concrete (up to LFC 30; LFC is the code name of strength grade of mortar). Because of these properties, it can be widely used in non-structure and structure concrete and pumped but not vibrated. The LWAs are SC and SP in foamed concrete in this study, and the mixes are shown in **Table 20**.

Although there are countless air pores in ALWFC, most of these pores are discontinuous, so ALWFC has better durability. According to the test results, the carbonation depth generally does not exceed 5 mm in 56 days, and the resistance performance with regard to chloride ion permeability, that is, the electric flux after 6 hours, is smaller than 1000 C. On the other hand, the ALWFC also has better fire resistance, sound insulation, and sound absorption capabilities.

6. Properties of reinforced ALWC

Taking a reinforced ALWC beam, for example, the parameters of the tested beams are shown in **Tables 21–25** and **Figures 5** and **6**.

The test results according to GB50152-2012 [15] are as follows.

According to [16, 17], during the beam flexural test, the maximum crack width should not exceed 0.3 mm under service loads, and the deflection should not exceed $l_0/200$ = 10.5 mm. The test values are 0.27 and 5.21 mm for the crack width and deflection, respectively, so ALWC can meet the code requirements. On the other hand, the crack load is about 28% of the ultimate load, and the service load is about 72%; these values are basically the same as those for the RC beam.

For the shear beam, because there is no warning before the occurrence of diagonal cracks, the diagonal cracks occur in the shear span section when the load is 20% of the ultimate load and then rapidly expand to the length of 100–150 mm. The initial width of the diagonal crack is generally 0.05 mm in the reinforced NWC beam compared to 0.03 mm in this study. At the same time, the maximum width of cracks and the deflection under service loads are 0.29 mm and 10.05 mm, respectively, thus meeting the code requirements.

The theoretical and test values of ultimate strength for normal and diagonal sections are shown in **Table 25**. All the test values slightly exceed the theoretical values. Compared to a reinforced NWC beam with the same stiffness, the width and height of the cross-section need to be increased by 18%, respectively. On the other

No.	$b \times h$ (mm)	h_0 (mm)	l_0 (mm)	λ	Longitudinal tension bars			Stirrup	
					Bar	A_s (mm^2)	ρ_s (%)	Bar	ρ_{sv} (%)
B1	150 × 300	267	2100	—	2$\underline{\Phi}$16	402	1.00	Φ8@100	0.67
B2	150 × 300	270	2100	—	2$\underline{\Phi}$10	157	0.39	Φ8@100	0.67
B3	150 × 300	264	2100	—	3$\underline{\Phi}$22	1140	2.89	Φ8@100	0.67
S4	150 × 300	267	2100	2	2$\underline{\Phi}$16	402	1.00	Φ8@140	0.48
S5	150 × 300	270	2100	2	2$\underline{\Phi}$10	157	0.39	Φ8@140	0.48
S6	150 × 300	265	2100	2	2$\underline{\Phi}$20	628	1.58	Φ8@140	0.48
S7	150 × 300	267	2100	0.95	2$\underline{\Phi}$16	402	1.00	Φ8@140	0.48
S8	150 × 300	267	2100	1.5	2$\underline{\Phi}$16	402	1.00	Φ8@140	0.48
S9	150 × 300	267	2100	3.05	2$\underline{\Phi}$16	402	1.00	Φ8@140	0.48
S10	150 × 300	267	2100	2	2$\underline{\Phi}$16	402	1.00	Φ8@140	0.48
S11	150 × 300	267	2100	2	2$\underline{\Phi}$16	402	1.00	Φ8@100	0.67
S12	150 × 300	267	2100	2	2$\underline{\Phi}$16	402	1.00	Φ8@180	0.37

Notes: (1) The notations 'B' and 'S' stand for bend and shear, respectively; (2) b and h stand for the width and height of a beam cross, respectively; (3) l_0 stands for the calculated span; (4) λ stands for the ratio of shear span to effective depth; (5) $\underline{\Phi}$ and Φ stand for hot rolled crescent-shaped bars (HRCSB, hereinafter referred to as crescent ribbed bars, CRB) and hot rolled plain steel bars (HRPSB, hereinafter referred to as plain steel bars, PSB), respectively; (6) A_s stands for transverse area; (7) ρ_s and ρ_{sv} stand for the ratio of reinforcement and ratio of stirrup reinforcement, respectively; (8) @ stands for the spacing between stirrups; (9) the diameter of a steel bar means the nominal diameter (d), and the concrete is ALWC with LC30; (10) each of the test beams has two hanger bars (2$\underline{\Phi}$12) in order to meet detailing requirements.

Table 21.
Parameters of test beams for bend and shear, respectively.

Type	d (mm)	f_y (MPa)	f_u (MPa)	E_s (MPa)
PSB	8	316	434	2.1×10^5
	10	329	457	2.0×10^5
	12	335	482	2.0×10^5
CRB	16	342	527	2.0×10^5
	18	362	576	2.0×10^5
	22	396	612	2.0×10^5

Notes: f_y stands for yield strength; f_u stands for ultimate tensile strength.

Table 22.
Parameters of PSB and CRB.

hand, if the section remains unchanged, according to the numerical simulation results for a seven-storey residential building, the total weight of the building is reduced by around 14.8%, and the inter-storey displacement angle is increased by around 27.5% under earthquake loading. This is because ALWC has a bigger ratio of cubic compressive strength to dry apparent density, a smaller elastic modulus, and a larger anti-deformation capacity.

No.	Cracking load and maximum width of crack		Service load and maximum width of crack		Failure load and maximum width of crack	
	P_{cr} (kN)	$\omega_{cr, max}$ (mm)	P_k (kN)	$\omega_{k, max}$ (mm)	P_u (kN)	$\omega_{u, max}$ (mm)
B1	30	0.11	70	0.24	100	1.85
B2	15	0.15	35.7	0.27	51	1.54
S4	45	0.14	120.5	0.19	160	0.63
S5	30	0.01	130.9	0.22	187.1	0.43
S6	25	0.02	129.7	0.25	185.3	0.41
S7	45	0.01	145.3	0.28	207.5	0.57
S8	40	0.012	135.9	0.29	194.1	0.63
S9	30	0.013	119.5	0.26	170.7	0.50
S10	35	0.017	139.0	0.26	198.6	0.56
S11	42	0.019	166.3	0.21	237.5	0.49
S12	30	0.01	115.2	0.21	164.6	0.43

Table 23.
Maximum crack widths under different load stages.

No.	Yield load and deflection		Ultimate load and deflection	
	P_y (kN)	δ_y (mm)	P_u (kN)	δ_u (mm)
B1	85	5.21	100	29.6
B2	43.4	4.11	51	19.75
B3	—	—	160	5.63
S4	150.3	9.52	187.1	29.76
S5	140.5	9.51	185.3	29.53
S6	165.4	8.87	189.6	29.89
S7	180.6	9.02	207.5	23.87
S8	145.3	9.24	194.1	29.79
S9	115.8	10.05	170.7	27.82
S10	150.5	9.41	198.6	28.42
S11	175.9	9.53	237.5	29.92
S12	125.2	9.76	164.6	28.57

Table 24.
Deflections under yield and ultimate load, respectively.

No.	f_c (MPa)	f_y (MPa)	A_s (mm^2)		h_0 (mm)	b (mm)	M_u^c (kN·m)	M_u^t (kN·m)	M_u^t/M_u^c
B1	31.80	342	402		267	150	34.73	35	1.01
B2	30.51	329	157		270	150	13.65	17.85	1.30
No.	f_t (MPa)	f_y (MPa)	A_{sv} (mm^2)	λ	h_0 (mm)	b (mm)	V_{cs}^c (kN·m)	V_{cs}^t (kN·m)	V_{cs}^t/V_{cs}^c

S4	1.56	316	100.5	2	267	150	91.81	93.55	1.02
S5	1.50	316	100.5	2	270	150	91.62	92.65	1.01
S6	1.59	316	100.5	2	265	150	91.71	94.83	1.03
S7	1.64	316	100.5	0.95	267	150	111.09	103.75	0.93
S8	1.53	316	100.5	1.5	267	150	97.33	97.05	1.00
S9	1.58	316	100.5	3.05	267	150	84.00	85.35	1.02
S10	1.62	316	100.5	2	267	150	93.01	99.30	1.07
S11	1.54	316	100.5	2	267	150	115.63	118.75	1.03
S12	1.57	316	100.5	2	267	150	78.55	82.3	1.05

Notes: (1) f_t stands for axial tensile strength obtained by the test or calculated directly by splitting tensile strength or bending strength; (2) M_u^c and M_u^t stand for the ultimate bending moment of the normal section calculated by the code and the test values, respectively; (3) V_{cs}^c and V_{cs}^t stand for the ultimate shear strength of the diagonal section calculated by the code and the test values, respectively.

Table 25.
Theoretical and test values of ultimate strength for normal and diagonal sections, respectively.

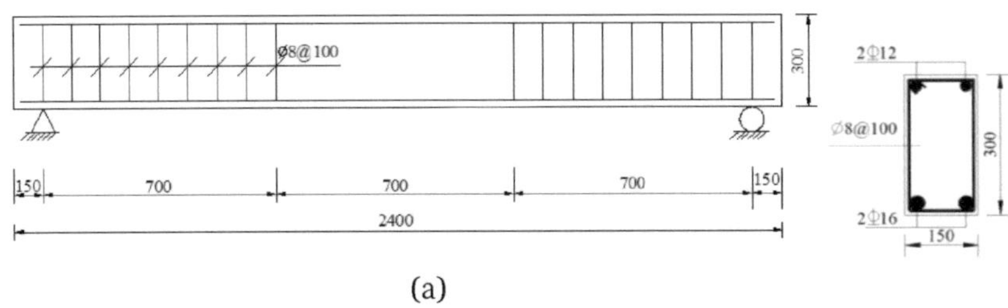

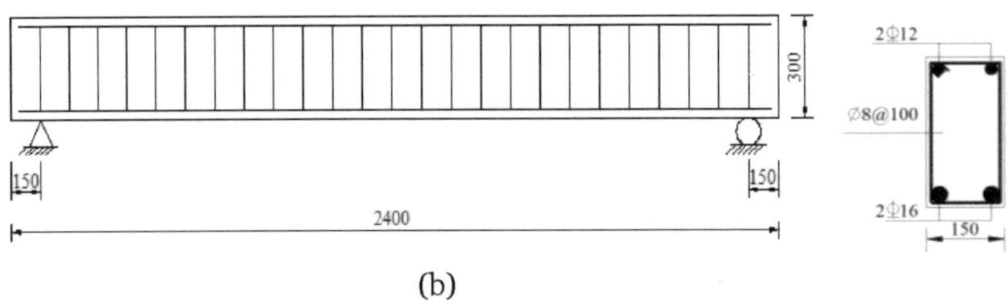

Figure 5.
Sketch of reinforcement for bending and shear beams, respectively. (a) Sketch of reinforcement for bending beam, and (b) Sketch of reinforcement for shear beam.

7. Properties of lightweight sand mortar

7.1 General performance of mortar

Because of the rough surface and higher porosity, SP can absorb cement particles and water, which leads to poor mixture workability, so an admixture of cellulose ether, emulsion powder, and so on must be added to meet the code requirements for the mortar consistency and delamination degree of mortar [18–20]. The mix proportions for different strength grades and the test results are shown in **Table 26**.

Compared with the code [18, 19], the dry apparent density of lightweight sand mortar is smaller than 1900 kg/m^3, and the heat conduction coefficient is smaller than 0.8–1.0 W/(m·K). Because of the porosity of SP, the dry apparent density and heat conduction coefficient are smaller than those of normal-weight sand mortar, so SP has better thermal insulation performance.

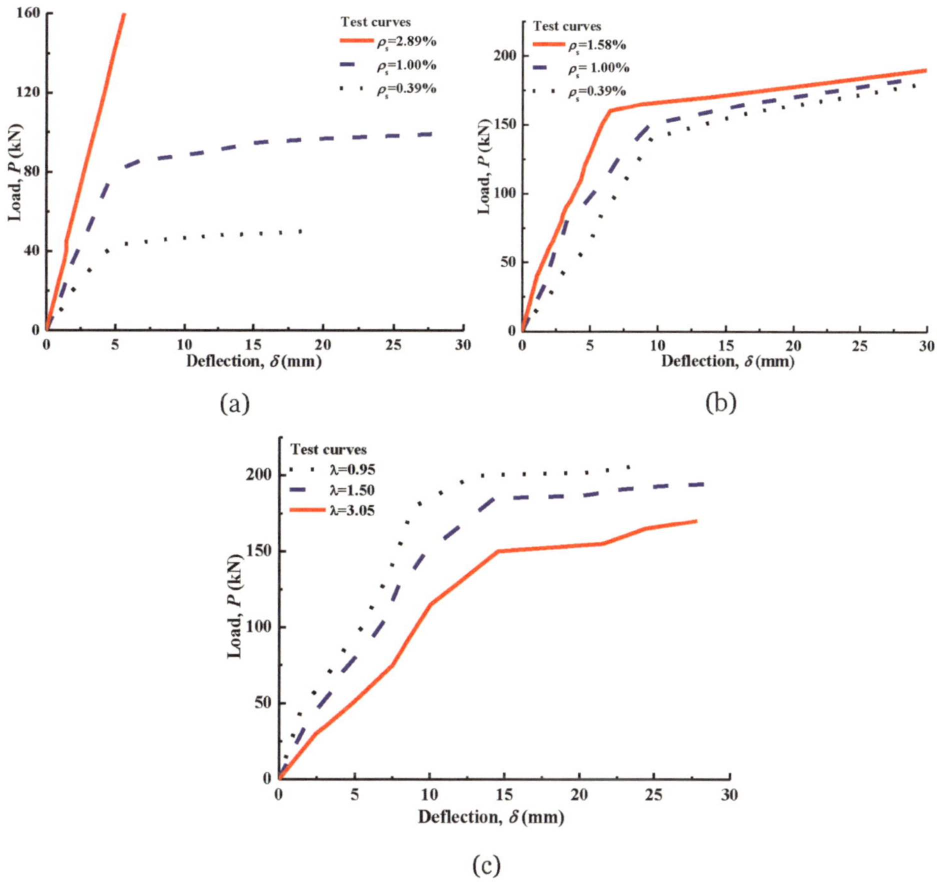

Figure 6.
Test curves for load deflection. (a) Bending beam, (b) Shear beam, and (c) Shear beam.

7.2 Durability of mortar

Taking carbonisation, chloride iron penetration, and sulphate attack, for example, the test results are shown in **Table 27**.

The durability of mortar is enhanced with increases in the strength grade. Especially until 30 cycles, the mass increases. Analogously, the sulphate resistance coefficient is also enhanced until 15 cycles. The reason for this is the porosity and water absorption capacity of SP, which can strengthen the internal curing capacity and thus promote the hydration reaction.

7.3 Fire-resistant performance of mortar

The test results for cubic compressive strength, tensile bond strength, and heat conduction coefficient after elevation of the temperature are shown in **Table 28**, where the average values of mass loss are 9, 19, 39, 53, and 65 g after 100, 200, 300, 400, and 500°C, respectively.

The behaviour of the mortar is similar to that of concrete after the elevation of temperature, and the residual strengths after high-temperature treatment are almost 75% at 500 °C and can therefore meet the fire protection design requirements. Below 300°C, the strength and heat conduction coefficient increase; however, at temperatures above 300°C, the strength and heat conduction coefficient decrease, and all of the parameters increase with increases in the strength grade.

Strength grade	m_C (kg)	m_{FA} (kg)	m_{SP} (kg)	ω_{EP} (%)	ω_{CE} (%)	m_W (kg)	*MC* (mm)	*DM* (mm)	f_{cu} (MPa)	f_{tb} (MPa)	λ ($w \cdot m^{-1} k^{-1}$)	ρ_d (kg/m^3)
LM 5.0	162	28	880	1.75	0.35	285	77	12	6.0	0.2319	0.2432	1132
LM 7.5	184	32				273	76	14	7.9	0.3163	0.2762	1217
LM 10	206	35				260	79	9	12.4	0.5135	0.2777	1188
LM 15	251	43				260	82	10	17.2	0.5802	0.2982	1273
LM 20	296	51				248	76	9	22.5	0.5781	0.3190	1348

Notes: (1) ω_{EP} and ω_{CE} are the emulsion powder (EP) and cellulose ether (CE) content (mass) of cement, respectively; (2) MC and DM stand for the mortar consistency (MC) and delamination degree of mortar (DM), respectively; (3) f_{tb} stands for the tensile bond strength; (4) λ stands for the heat conduction coefficient.

Table 26.
Reference mixes (1 m^3) and test results of lightweight sand mortar.

Strength grade	$h_c^{28\,d}$ (mm)	$Q_e^{56\,d}$ (C)	Mass loss (g)			Corrosion resistance coefficient (%)		
			Δm^{15}	Δm^{30}	Δm^{60}	K_f^{15}	K_f^{30}	K_f^{60}
LM 5.0	19.42	1042.3	18.1	20.2	−12.1	130	78.4	40.8
LM 7.5	18.32	902.4	16.3	19.6	−10.9	128	79.8	55.7
LM 10	18.01	858.4	16.5	17.7	−9.4	125	83.9	68.4
LM 15	17.65	743.8	15.5	16.9	−8.5	120	85.3	71.1
LM 20	15.13	701.2	13.8	15.1	−7.7	119	91.4	75.6

Table 27.
Test results for durability of lightweight sand mortar.

	T (°C)	LM 5.0	LM 7.5	LM 10	LM 15	LM 20
f_{cu} (MPa)	100	6.5	8.3	12.9	18.0	23.4
	200	7.1	9.2	13.8	18.7	24.1
	300	7.8	10.3	14.9	18.9	24.7
	400	5.7	7.6	11.2	16.0	20.4
	500	4.7	6.3	9.2	14.3	18.7
η_{cu}^{T} (%)	500	78.3	79.7	74.2	83.1	83.1
f_{tb} (MPa)	100	0.100	0.146	0.349	0.366	0.397
	200	0.114	0.160	0.420	0.468	0.510
	300	0.123	0.175	0.430	0.504	0.532
	400	0.090	0.117	0.308	0.328	0.348
	500	0.083	0.106	0.170	0.214	0.239
λ (w·m^{-1} k^{-1})	100	0.231	0.253	0.260	0.275	0.299
	200	0.211	0.239	0.248	0.259	0.277
	300	0.220	0.241	0.253	0.263	0.281
	400	0.205	0.226	0.230	0.241	0.259
	500	0.204	0.218	0.222	0.236	0.254

Table 28.
Strength and heat conduction coefficient of lightweight sand mortar after elevated temperature treatment.

8. Summary

ALWC has a number of advantages and disadvantages. Using NS (or MS) and crushing stone to replace a part of LWAs alone or at the same time in equal volume ratio, the new concrete types can be called semi-lightweight concrete (semi-LWC), which includes SLWC, GLWC, HALWC, and so on. Semi-LWC can not only reduce the cost of ALWC but also increase the properties of ALWC, such as workability, strength, durability, anti-deformation, fire resistance, and so on. Especially, moderate amounts of mineral powder and limestone powder can significantly increase the strength and durability.

All types of the concrete can meet the Chinese National Code requirements as well as have a smaller heat conduction coefficient and higher ratio of cubic compressive strength to dry apparent density than NWC. However, the effect of NWAs on semi-LWC is different. Gravel aggregates are bigger than sand aggregates, so the effect is more complex when added simultaneously. At the same time, the multiaxial strength increases with increasing lateral pressure, and the ratio of biaxial compressive strength to uniaxial compressive strength is slightly larger compared to NWC. This is because the NWA has better thermal conductivity and a small quantity of NWAs can help to reduce autogenous shrinkage of ALWC. On the other hand, the axial compressive strength of ALWC is close to cubic compressive strength, which shows that the ALWC has a self-lubricated antifriction effect because of the SC. On the contrary, if the diameter of SC is too large, its TCS is too small. The strength of ALWC decreases with increasing curing age, so the diameter of SC should not exceed 15 mm in general.

Although the SC has a softening effect, the LWACs do not, so they can be used in hydraulic structure engineering. Even the stiffness of the reinforced ALWC is smaller than that of NWC; because of the smaller modulus of elasticity and apparent density, the reinforced ALWC has a better bending and shear properties. Moreover, the maximum width of crack in ALWC is smaller than that in NWC, so buildings made of ALWC can have better anti-seismic properties.

The highest strength grade of foamed concrete made of SC and SP can be up to LFC 30, so it can be used in non-structure and structure construction, and it has better performance in terms of thermal resistance, sound absorption, insulation, fire resistance, and so on. On the other hand, mortar made of SP can also be used in plastering mortar and masonry mortar and has the abovementioned excellent characteristics.

At high temperature, the performances of LWACs and mortar decrease with increasing temperature but can be increased with increasing lateral pressure. In negative-temperature curing within the range of −15°C, the LWACs can meet the construction requirements of the artificial freezing method.

Finally, it should be pointed out that the descending curve of the stress-strain curve of LWACs cannot be measured easily, especially after elevation of the temperature. Even so, the formulas provided in this paper can meet the demands of experimental precision under both uni- and multiaxial stress states.

Acknowledgements

This work was financially supported by the National Natural Science Foundation of China (41172317; 51774112; 51474188). This chapter was polished by Proof-Reading-Service.com Ltd.

Author details

Jianhui Yang
Henan Province Engineering Laboratory for Eco-Architecture and the Built Environment, Henan Polytechnic University, Jiaozuo, PR China

*Address all correspondence to: yangjianhui@hpu.edu.cn

References

[1] GB/T 14684–2011. Sand for Construction. China; 2011 (in Chinese)

[2] GB/T 17431.1-2010. Lightweight Aggregates and its Test Methods—Part 1: Lightweight Aggregates. China; 2010 (in Chinese)

[3] JGJ/T 241-2011. Technical Specification for Application of Manufactured Sand Concrete. China; 2011 (in Chinese)

[4] JGJ 52-2006. Standard for Technical Requirements and Test Method of Sand and Crushed Stone (or Gravel) for Ordinary Concrete. China; 2010 (in Chinese)

[5] GB/T 50266-2013. Standard for Test Methods of Engineering Rock Mass. China; 2013 (in Chinese)

[6] GB 50496-2009. Code for Construction of Mass Concrete. China; 2009 (in Chinese)

[7] Taylor MA, Jain AK, Ramey MR. Path dependent biaxial compressive testing of an all-lightweight aggregate concrete. Automation (Cleveland). 1972;**69**(12): 758-764

[8] JGJ 51-2002. Technical Specification for Lightweight Aggregate Concrete. China; 2002 (in Chinese)

[9] GB/T 17431.2-2010. Lightweight Aggregates and its Test Methods – Part 2: Test Methods for Lightweight Aggregates. China; 2010 (in Chinese)

[10] GB/T 50082-2009. Standard for Test Methods of Long-Term Performance and Durability of Ordinary Concrete. China; 2009 (in Chinese)

[11] Guo Z, Zhang X, Zhang D, et al. Experimental investigation of the complete stress-strain curve of concrete. Journal of Building Structures. 1982; **3**(1):1-12 (in Chinese)

[12] Yang J, Yang Z, Huang H, et al. Multiaxial strength model of concrete. Engineering Mechanics. 2008;**25**(11): 100-110 (in Chinese)

[13] JG/T 266-2011. Foamed concrete. China; 2011 (in Chinese)

[14] JGJ/T 341-2014. Technical specification for application of foamed concrete. China; 2014 (in Chinese)

[15] GB/T 50152-2012. Standard for test method of concrete structures. China; 2012 (in Chinese)

[16] JGJ 12-2006. Technical specification for lightweight aggregate concrete structures. China; 2006 (in Chinese)

[17] GB 50010-2010. Code for design of concrete structures. China; 2010 (in Chinese)

[18] JGJ/T 98–2010. Specification for mix proportion design of masonry mortar. China; 2010 (in Chinese)

[19] JGJ/T 220-2010. Technical specification for plastering mortar. China; 2010 (in Chinese)

[20] JGJ/T 70-2009. Stand for test method of basic properties of construction mortar. China; 2009 (in Chinese)

Section 4

Bio-Cemented Sands

Chapter 6

Geomechanical Behavior of Bio-Cemented Sand for Foundation Works

Youventharan Duraisamy and David Airey

Abstract

Bio-cementation is an innovative green technology that complements existing ground improvement techniques, but it is yet to be proven for large-scale foundation works. Previously attention has been focused on strategies to inject the bacteria and nutrients to produce the cement in the ground. This study looks at the performance of geomechanical response when the bacteria and nutrients are mixed in sand, an approach that is used in producing cemented soil columns. To explore the mechanical response of bio-cemented soil, results from unconfined compressive strength (UCS) tests and triaxial tests have been analyzed to understand the effects of bio-cementation for sand in contrast to alternative cement, gypsum. The stiffness has also been monitored using bender element techniques in triaxial cell. Both the shear wave signals during the cementation phase and the shearing phase were recorded using this technique. The results show that for a given amount of cement, higher resistances are measured for the bio-cemented samples compared to gypsum. The mixing process is shown to produce homogeneous bio-cemented samples with higher strength and stiffness than the technique of flushing or injection commonly used, provided the amount of calcite is less than 4%. The results show that the bio-cement produces similar mechanical behavior to other artificially cemented sands.

Keywords: bio-cement, calcite precipitation, Sydney sand, strength, stiffness, shear wave velocity, small strain modulus, urea hydrolysis

1. Introduction

The study on the geomechanical behavior of bio-cemented Sydney sand was carried out using unconfined compression strength (UCS) tests and triaxial tests. To date there have been limited numbers of triaxial tests reported on soil cemented by microbes, and owing to differences in soil type and cementation methodology, these have provided variable results on the potential of bio-cementation. Part of the variability in behavior is a result of the widely used method of injecting bacterial solutions into sand to create the cementation. The injection process leads to heterogeneity in the distribution of calcite and hence to variability in the mechanical response and permeability of the bio-cemented soil. The mixing technique was used to create consistent and coherent bio-cemented samples and to overcome the difficulties of interpretation faced by previous researchers [1–3]. It has been suggested that the injection technique targets the contact points between the particles, which is beneficial for cementing, and comparisons are made between mixed samples in this study and other studies where injection has been used to explore this hypothesis. The small strain shear stiffness has been used as an indicator to evaluate the success of bio-cementation [4], and shear wave velocity was monitored in this study to explore the influences of cement level and stress on the data and the ability to predict the success of bio-cementation.

The discussion in this chapter concentrates on:

i. The effect of preparation and curing time on cementation level and corresponding amount of calcite/cement

ii. A comparison of UCS strengths of bio-cemented and gypsum-cemented specimens

iii. The effects of confining stress of soil on the geomechanical behavior of bio-cemented specimens

iv. The influence of calcite content on the strength and stiffness

2. Literature review

Over the past decade, the potential for microbially induced calcite precipitation (MICP), or simply bio-cement, to improve soil and rock responses has been extensively studied by petroleum, geological, and civil engineers [4, 5]. Recently, studies were undertaken to understand the geomechanical parameters of granular soils using microbes in biochemical process, which produce bio-cement in the subsurface [1, 6, 7]. It has been suggested [8] that these reactions simulate the natural geochemical processes that transform sand into sandstone. However, the MICP process is rapid and produces a precipitate with soft and powdery crystals, whereas natural limestone forms slowly and creates a very hard precipitate [6]. Most of this research has focused on the use of ureolytic bacteria, which have been shown to be capable, with the addition of urea and reagents, of producing calcite that binds to soil particles [6–13].

The general trends of cementation effects on granular material are increases in strength and stiffness, which increase with the amount of cementing material, although this may vary greatly depending on the amount of cementing material used. It has been noted that the effectiveness of cement depends on the density, the effect of cement being greater at lower densities [14–17]. Many studies on artificially cemented soils have shown that cementation significantly increases the initial tangent modulus of a soil and monitoring the stiffness can be a useful method of tracking the amount of cementation [18, 19]. A range of cementing agents have been investigated including ordinary Portland cement (OPC), gypsum, sodium silicates, and calcium carbonate [14–18] to understand the influence of cementation and to simulate materials used in ground improvement work. Generally, the geomechanical responses of bio-cemented granular soil are similar to any other artificially cemented granular soil [7].

Although the cementing effect is more significant in loose sand, it is found that more bio-cement is needed to achieve the strength of dense sand when applied in loose sand [13]. It has also been suggested that the growth of calcite crystals at points of contact between sand grains has a significant influence on UCS strength [20]. As bacteria and nutrients are pumped through sand continuing calcite precipitation, it can lead to a large proportion of the voids being filled, and high

UCS can be obtained. For example, strengths of up to 30 MPa [20] were obtained from small-scale experiments, and even at larger-scale (100 m^3) strengths of up to 12 MPa [21] have been reported. Other results show maximum compressive strengths obtained for bio-cemented sand of about 14 MPa [22]. As much as UCS strength increase can be related with individual soil particle strength, factors like roundness, size, and shape too may reduce the strength [20]. An increase in shear strength of 35, 50% and more than 100% is observed for round coarse particles, coarse angular particles, and round fine particles, respectively [23]. According to Al Qabany and Soga [24], for the same amount of precipitated calcite, the greatest UCS strength was obtained when the concentration of the solution was low. These results were obtained over a range of different initial relative densities. Cheng et al. [17] also reported that it is possible to get higher UCS strengths at the initial phase when sample is low or partially saturated. The different soils used and different preparation procedures have resulted in a wide range of parameters to describe the bio-cemented soils. For example, clean Ottawa sand treated with microbes produced calcite in the range of 0–4%. During the treatment, the angle of friction increased from 35.3 to 39.6° and the cohesion from 0 to 93 kPa [25].

A range of applications for bio-cement have been suggested. A recent example is a feasibility study carried using bio-cement for slope stabilization by means of a surficial treatment where a layer of hard stratum was obtained on the subsurface with UCS strength of 420 kPa after 10 days of treatment [7, 26]. However, this method of treatment is not extensively applied in the field of ground engineering.

3. Materials and methods

Microbes, chemical substrate, and reagents have been used to produce bio-cemented specimens. This process uses bacteria to catalyze the urea hydrolysis reaction that precipitates calcite. The ureolytic bacteria known as *B. megaterium* (strain ATCC 14581) were used. To produce sufficient bacteria for the cemented samples, a KWIK-STIK (produced by Microbiologics®) containing the microorganism strain was cultured in batches using liquid medium. The growth medium (refer to **Table 1**) in liquid form was prepared in advance and placed in the incubator at 30°C for 24 hours using a 50 ml beaker. Importantly, this bacterium is nonpathogenic and poses no harm to humans.

Fixed quantity of clean, dried sand was placed in a mixing bowl. Then the required amount of urea powder and calcium chloride powder was added based on a percentage of the sand weight free from moisture. The nutrient masses (urea and calcium chloride) ranged from 5 to 20% of the sand weight free from moisture. Additional water was added to facilitate mixing to give a water mass of about 10% of the mass of the dry ingredients. The ingredients were then thoroughly mixed for

Ingredients (L^{-1})	
Nutrient broth	3 g
Urea	20 g
NH_4Cl	10 g
$NaHCO_3$	2.12 g
$CaCl_2$	2.8 g

Table 1.
Typical liquid medium or broth.

1 min before being placed in cylindrical molds. To prepare uniform and reproducible samples with a consistent density, the mixture was divided into five portions before being filled in the molds and gently tamped each time.

The bio-cemented samples have been compared with gypsum-cemented samples which have been prepared by combining the dry ingredients (sand and unhydrated gypsum) followed by mixing with water and placing in a mold similarly to the bio-cemented samples.

The preparation technique produced cylindrical samples with 55 x 110 mm in dimensions. After extraction and curing, the samples were either placed directly in a compression machine to perform UCS tests or in a fully computerized triaxial testing apparatus to perform geomechanical tests with elevated confining stresses, which was also fitted with bender elements to monitor the secondary (shear) wave pulse. Once the UCS and triaxial shearing tests were completed, bio-cemented samples were extracted and analyzed to determine the amount and distribution of the calcite precipitated. Further details of methods and procedures are provided by Duraisamy [2].

4. Unconfined compressive strength tests

Strength tests have been performed using Sydney sand mixed with two cementing media, gypsum and the bacterial mixture. **Figure 1** shows the responses of the UCS tests of cemented sand performed with gypsum contents in the range of 5–20%. All samples fail at small strains with brittle responses. A clear trend of increasing strength with gypsum content is evident. However, interpretation of these tests requires caution as strength is affected by density, which tends to increase as fines are added; the limited water retention in clean sand, which limits the water available for hydration; and the presence of suctions in the tested samples (dried in laboratory conditions), which tends to increase the strength. All these factors tend to enhance the effectiveness of the cement as more cement is added, at least initially.

Bio-cemented samples of Sydney sand and its UCS strength are shown in **Figure 2**. As expected, as the amount of calcite precipitation increases, the UCS readings also increases. Similarly to the gypsum-cemented samples, the bio-cemented sand samples show generally stiff and brittle behavior, although with more ductility than for the gypsum cement. The results accord with other

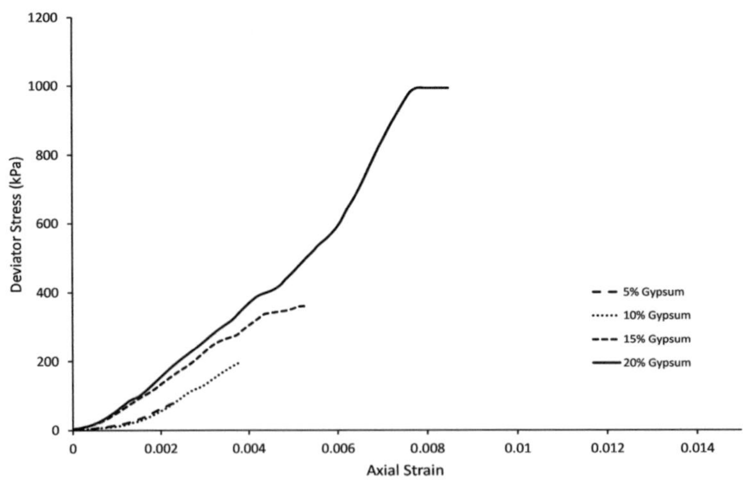

Figure 1.
UCS test responses from gypsum-cemented specimens.

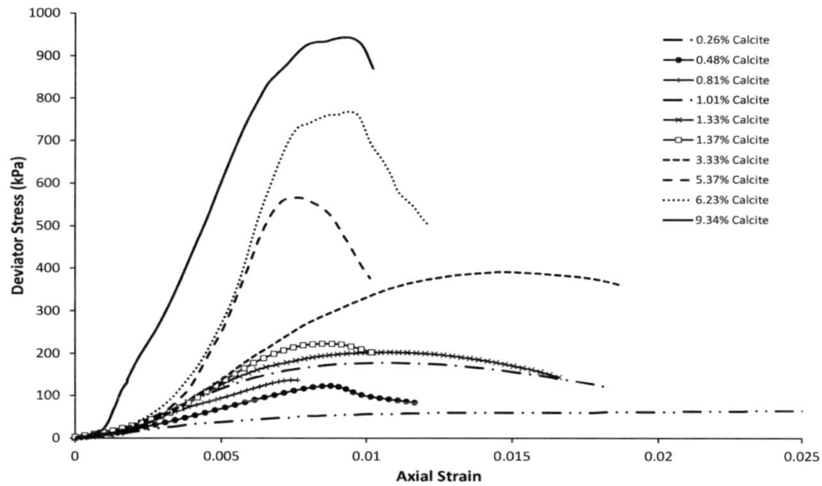

Figure 2.
UCS responses for bio-cemented specimens.

Test No/ID	Test type	Average calcite (%)	Top (%)	Middle (%)	Bottom (%)	Standard variance (±%)	Sample variance (%)
3%B	UCS	1.33	1.52	1.14	1.34	0.19	3.6
B14	Triaxial	1.54	1.49	1.51	1.61	0.10	1.1
5%B	UCS	2.73	2.73	2.91	2.68	0.12	1.5
B10	Triaxial	2.61	2.94	2.79	2.79	0.16	2.7
10%B	UCS	5.33	5.52	5.21	5.26	0.15	2.3
B13	Triaxial	4.26	4.48	4.17	3.62	0.16	2.5
15%B	UCS	6.23	6.13	6.25	6.31	0.15	2.3
B16	Triaxial	6.98	7.09	6.90	6.91	0.11	1.1

Table 2.
Variance of calcite distributions in UCS and triaxial test samples.

studies [4, 7, 19], which have reported bio-cemented sand responds similarly to naturally and artificially cemented sand at low confining pressure. Comparison of **Figures 1** and **2** shows that samples with calcite contents less than 3.33% have lower stiffnesses than gypsum-cemented samples of the same strength. With low calcite contents, it is possible that sample heterogeneity influences the results, as reported in other studies [24] where the bio-solution has been pumped into the samples. However, as shown in **Table 2**, the mixing procedure used in this study has produced uniform calcite distributions through the samples, with less than 5% variance in different sample sections, for all calcite contents. End effects may also have affected the apparent stiffness as the sample ends were not prepared perfectly square and there was a tendency for water and bio-solution to flow out of the samples, because of the low water retention, during sample preparation. Nevertheless, the consistent trend with calcite content suggests that the results reasonably represent bio-cemented sand performance.

Figure 3 shows a comparison between the UCS responses of gypsum-cemented samples and bio-cemented samples of approximately similar strengths. It is evident that based on strength, also shown in **Figure 4**, the calcite produced by MICP is a much more effective cementing agent than gypsum, with about 0.5% of calcite

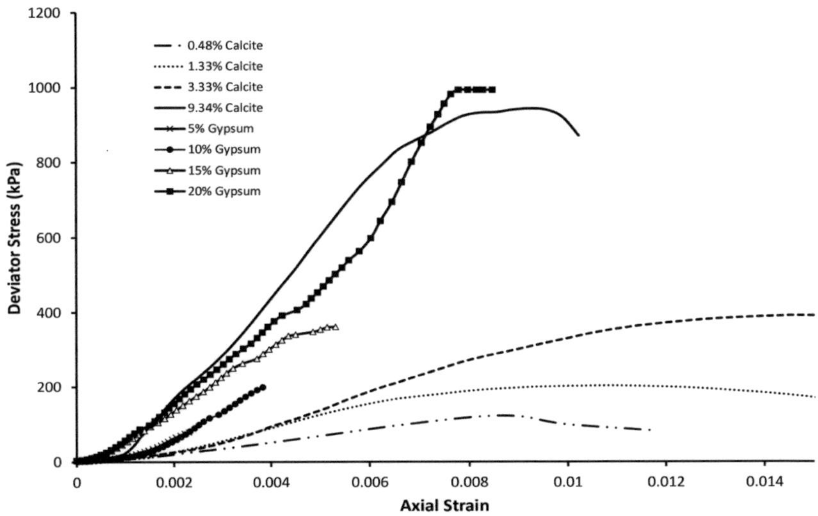

Figure 3.
Comparison of UCS responses of gypsum and bio-cemented specimens.

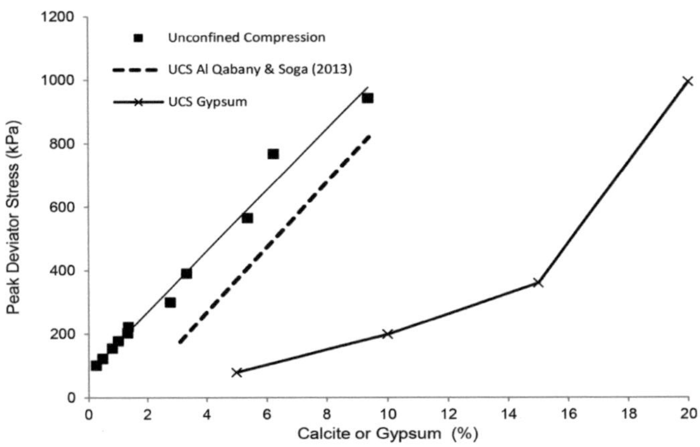

Figure 4.
Effects of cement content on UCS strength.

equivalent to about 5% of gypsum and 9% of calcite equivalent to 20% of gypsum. As noted above the stiffness and ductility of the bio-cemented samples are lower than for gypsum when cement contents are low, and this can in part be explained by the very low amounts of calcite required, and as shown in other studies [13, 27], much of this acts as space filler and does not actively contribute to the strength.

The influence of gypsum and bio-cement content on the strength improvement is shown in **Figure 4**. Previous report [24] also indicates similar UCS and calcite content relationship. However, the results from the current study in which mixing was used all lie above the previous research [24] in which the bio-cement solution was pumped into the samples. It was also reported [24] that preparing samples with low calcite contents by pumping in the solution was problematic as samples tended to have poor homogeneity and these weakly cemented samples tended to deform locally, giving low shear strength, and on occasion to collapse immediately upon loading. In contrast, the low calcite content bio-cemented specimens prepared by mixing in this study all showed significant improvements in resistance. As sands with similar gradings were used and samples were prepared to similar densities in both studies, the different responses point to the sample preparation method as being the key difference. Several studies [13–27] have shown that regardless of the injection process, obtaining uniform calcite precipitation is difficult, and it is also difficult to control, especially for small amounts of cement. The results from the mixing method of preparation show that this can lead to more homogenous cementation at low calcite contents and mixing can achieve calcite contents of nearly 10%. Nevertheless, the process of injecting bio-cement has advantages. It has been shown to be practical at field scale, and by using a series of injection phases, the process is capable of achieving very high strengths [27].

The strength and stiffness produced by the mixing technique varies depending on the soil and cement. Even though gypsum was mixed in with the soil, much more gypsum was required to produce the same strength and stiffness as the calcite cement. Because of its acicular particles, gypsum does not easily bridge between the sand particles, and it tends to fill the void spaces. Many studies have shown that in adding silt-sized fines, such as gypsum, fines fill the voids up to a transition fines content of approximately 25% after which the fines have increasing influence on the behavior. Once enough gypsum is present, it will fill the voids and form a strongly cemented matrix. Gypsum contents >15% appear to be needed to achieve this effect as illustrated in **Figure 4**.

5. Curing process and bender elements

The progress of the cementation process during curing was monitored using bender elements by recording the shear wave velocity change over time. This was achieved during the preparation of samples for triaxial testing. Split mold using PVC material fitted with a rubber membrane inside was designed to produce identical cylindrical samples with length of two times the diameter. Bender elements in the end platens transmitted waves vertically through the samples, and the waveforms and travel times were monitored using a semiautomated procedure [28]. The typical responses in **Figure 5** happen during the curing of gypsum-cemented samples. The hardening process occurs rapidly for gypsum contents above 10% and that curing is essentially complete after 1 h. This is consistent with the setting time reported by the gypsum supplier of 55 minutes. However, for low gypsum contents, there is some variability and longer setting times have been recorded. This is believed to be because

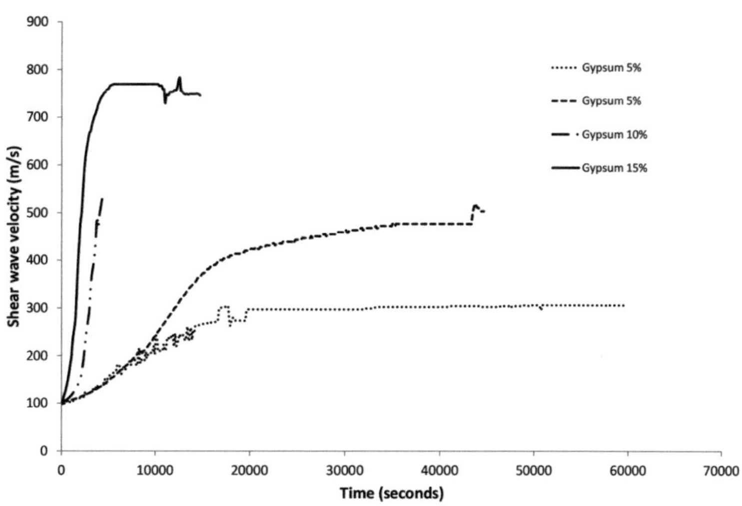

Figure 5.
Shear wave velocity changes during curing of gypsum-cemented specimens.

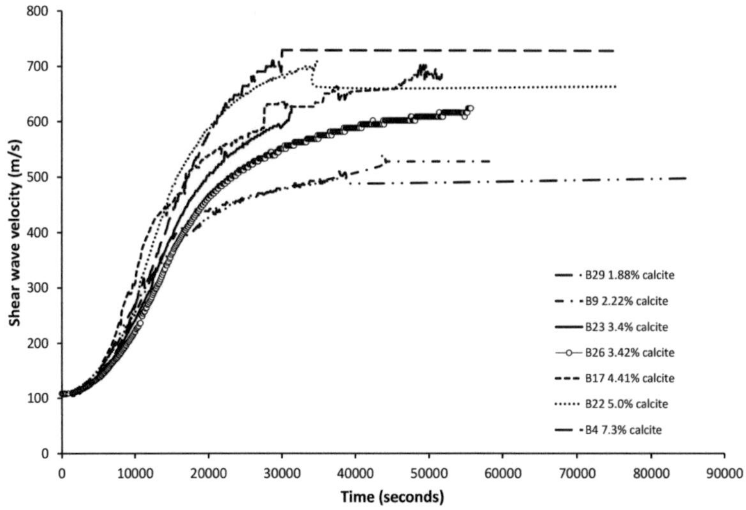

Figure 6.
Shear wave velocity changes during curing of bio-cemented specimens.

the setting of gypsum involves a hydration reaction that can be affected by changes in temperature and humidity. The samples were effectively sealed with limited supply of oxygen during the curing reaction, and with low gypsum contents, temperature rises associated with the exothermic reaction would be limited, hence limiting the reaction rate. After the hardening phase, the samples were left unattended overnight, and during this time no significant change in shear wave velocity was captured.

The variations of shear wave velocity during the calcite precipitation and hardening of the bio-cemented samples are shown in **Figure 6**. Small step changes in the shear wave velocity shown in **Figure 6** are a consequence of manual intervention in the semiautomated interpretation procedure and do not reflect the material response. For the range of final calcite values shown, the reaction time is very similar. In all cases there is a lag of about 1 hour before the cementation process begins, and the process is complete in about 12 hours after which the shear wave velocity remains constant. Samples were left for 24 hours before commencing the triaxial tests, and during this time the stiffness remained essentially constant. **Figure 6** also shows that the initial 100 m/s value increases over time, which is proportional to the stiffness and tends to increase with the calcite content as expected.

The comparative study on shear wave signal responses during curing for selected gypsum-cemented and bio-cemented sand samples is projected in **Figure 7**. The trend in the responses are similar for 5% gypsum and 1.88% calcite, whereas UCS tests have shown that the strength associated with 1.88% calcite is equivalent to about 10% gypsum. However, the rates of the cementation reactions depend on the chemistry of the hydration and MICP processes and are not expected to influence strength and stiffness. Nevertheless, it may be noted that the ratio between strength and stiffness varies with the cement type. The calcite-cemented samples have lower stiffness (shear wave velocity) than gypsum samples with the same strength. A likewise pattern was seen in the UCS tests and was inferred to be a simple consequence of the low amount of calcite cement.

No triaxial tests were performed with calcite contents lower than 1.88%, and thus it is unclear whether with lower calcite contents the reaction time will increase, which occurred for low gypsum contents. In other tests [19, 29], when the cementation occurred underwater, the time required for curing was greater than 24 hours, and it is expected that the curing time will depend on the chemical and

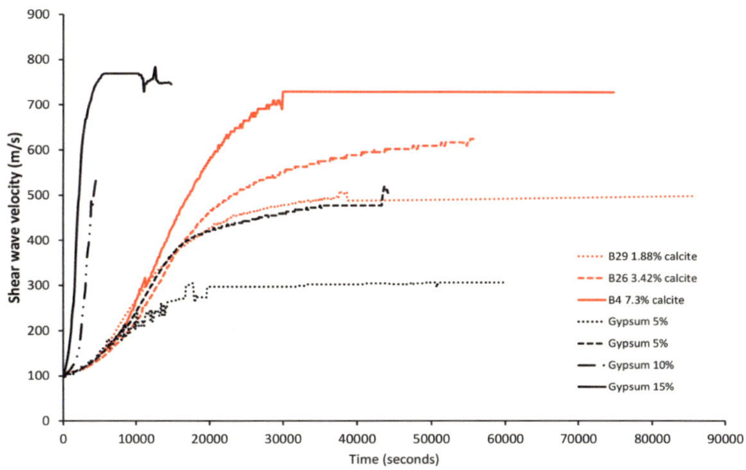

Figure 7.
Comparison of curing for gypsum and bio-cemented specimens.

environmental conditions. It may also be noted that the lag at the start of the cementation process is beneficial for both injection and mixing approaches at field scale.

6. Triaxial stress: strain responses

6.1 Uncemented sand

To enable the influence of the cement to be appreciated, triaxial tests have been performed on loose Sydney sand. Samples with various relative densities have been subjected to standard drained (CID) and undrained (CIU) tests with different confining pressures. Test results as in **Figure 8** are presented in terms of stress ratio (q/p′) against the axial strain. The results indicate that in all tests the stress ratio rises to a peak before gradually reducing toward a critical state value at large strain. Where the stress ratio dropped rapidly post peak, the samples had formed pronounced shear planes. The dotted line in **Figure 8** shows the estimated critical state stress ratio, M = 1.35, which corresponds to a friction angle of 32°.

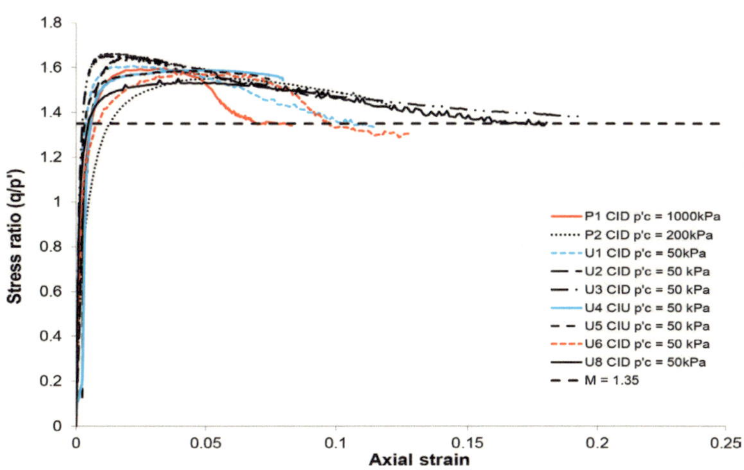

Figure 8.
Response of uncemented Sydney sand.

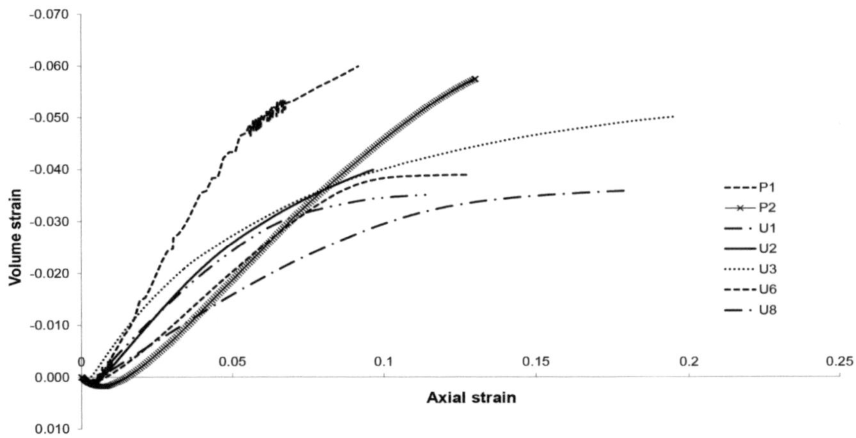

Figure 9.
Volumetric strains for CID tests on uncemented sand.

The volumetric responses from the drained tests reported in **Figure 8** are displayed in **Figure 9**. In all cases the samples expanded on shearing, which is consistent with their mobilizing peak stress ratios greater than the critical state value [30].

The bender element technique was used to obtain the shear wave velocity (V_s) for the uncemented sand. This has also allowed comparison with data on uncemented sand obtained from other studies and hence to demonstrate the reliability of the estimated soil stiffnesses. Knowing the shear wave velocity and bulk density (ρ) of the sand, the small strain shear modulus (G_{max}) can be determined from Eq. (1):

$$G_{max} = \rho V_s^2 \tag{1}$$

The parameter functions of G_{max} and the mean effective stress (p') for two typical uncemented sand samples, P1 and P2, are shown in **Figure 10(a)**. During isotropic compression an identical response is obtained, which may be described by Eq. (2):

$$G_{max} = 11.27\, p'^{0.475} \tag{2}$$

where G is in MPa and p' is in kPa.

Data on which Eq. (2) is based covers a range of p' from 10 to 3000 kPa, which is greater than incorporated in most published relations. For comparison, the data obtained in this study are plotted in **Figure 10(b)** with another published empirical relation for G_{max} which is given by Eq. (3). This incorporates a function of void ratio $f(e) = (2.17\text{-}e)^2/(1 + e)$ and constants A and n which are coefficients that depend on the type of material (**Table 3**).

$$G_{max} = A f(e)\, p'^{n} \tag{3}$$

The predicted G_{max} values from [31, 32] are similar to the Sydney sand data, and the linear relationship between G_{max} and p' in this study is closest to the equation proposed for Toyoura sand [32] which is to be expected given the similarity of mineralogy, particle size, and shape of the two granular materials.

6.2 Gypsum-cemented sand results

Results from triaxial tests on gypsum-cemented samples are presented in **Figures 11–13. Figure 11** shows the stress ratio and axial strain responses of cemented samples with gypsum contents between 5% and 20% and includes undrained and

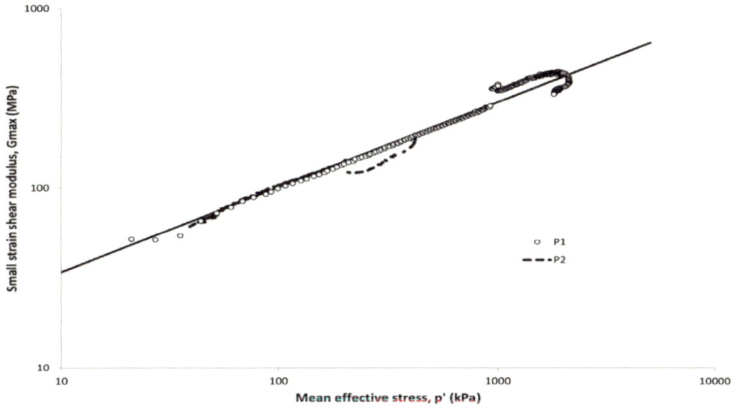

(a)

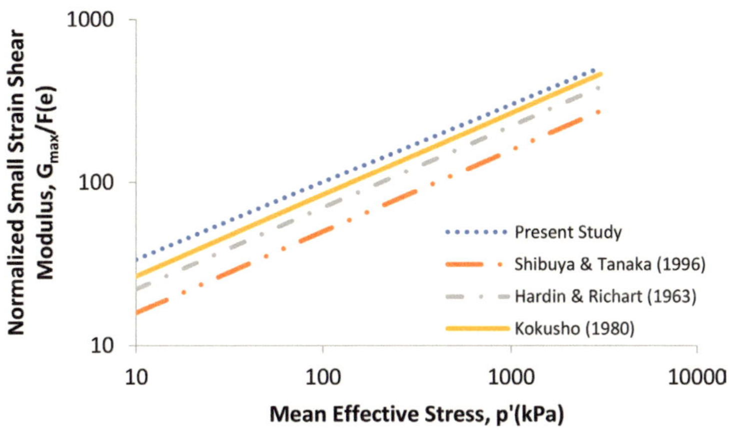

(b)

Figure 10.
Variation of G_{max} with p′ for uncemented sand (a) dry and saturated Sydney sand (b) validation with published data.

A	*n*	Reference
5000	0.5	Shibuya and Tanaka [31]
8400	0.5	Kokusho [32]
7000	0.5	Hardin and Richart [33]

Table 3.
Constants proposed for empirical equation G_{max}.

drained test results. Comparison with **Figure 8** for the uncemented sand shows the cemented samples developed high peak stress ratios and these are mobilized at lower axial strains than for the uncemented sand. For the drained tests, samples with confining stress of 50 kPa, the peak stress ratio and deviator stress increase with cement content as expected. For the undrained tests, a similar trend with cement content is apparent; however, for the more cemented samples, the stress ratio reaches the limiting value in the triaxial apparatus, which is 3. Once this occurs further loading is equivalent to performing a UCS test and the failure strengths of these samples are between 650 and 1300 kPa, in the range of the UCS strengths of the gypsum-cemented samples shown in **Figure 4**. After the peak the stress ratio reduces

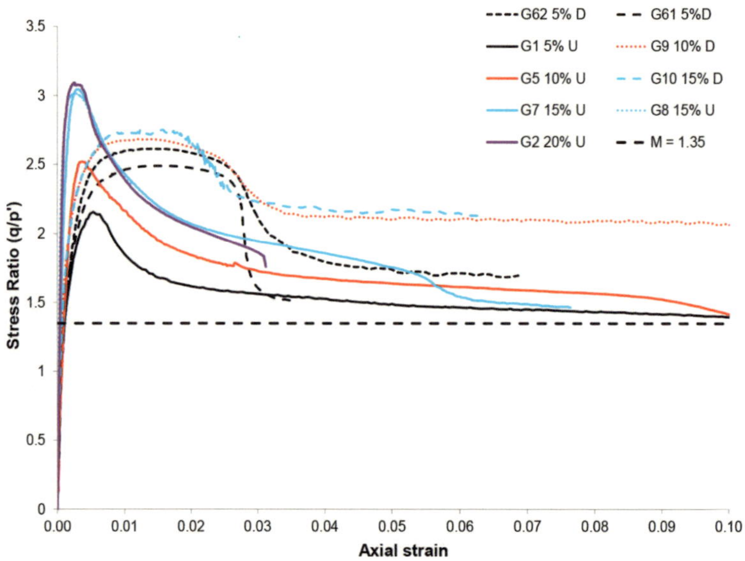

Figure 11.
Stress ratio, axial strain responses for all gypsum-cemented specimens.

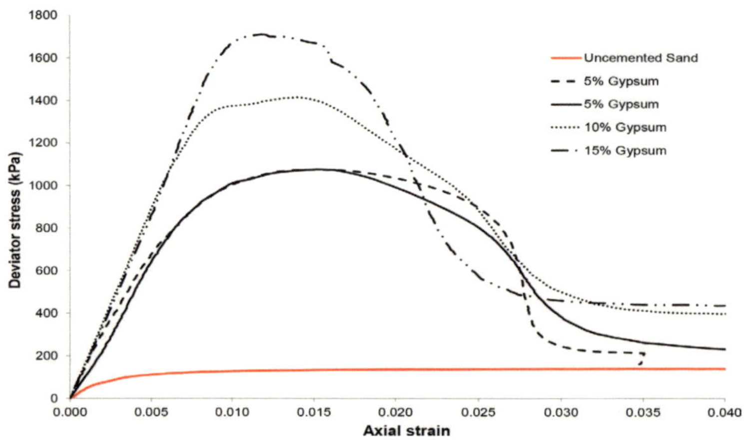

Figure 12.
Deviator stress, axial strain responses from drained tests (p'_c = 50 kPa).

approaching a constant value at large strain, however unlike the uncemented samples, the stress ratio does not appear to approach a unique value. It is believed that this is a consequence of nonhomogeneous deformation and if the samples could be sheared uniformly, they would approach the value of the ultimate critical state, M = 1.35 similar to the uncemented sand. Other studies (e.g., see [14, 34]) in which gypsum-cemented sands were tested showed that the presence of gypsum, up to 20%, did not influence the ultimate frictional resistance.

Figures 12 and **13** show the reactions of gypsum-cemented samples in drained triaxial tests. The gypsum cement leads to significant increases in strength compared with the uncemented sand, and even small amounts of gypsum increase the strength considerably. Further, the comparison with the UCS test responses shown in **Figure 1** indicates that there is a remarkable contribution with even a slight amount of gypsum, which is much greater in the triaxial tests. This is believed to be due to the applied confining stress which prevents the tensile failure mode that occurs in UCS tests. Samples cemented with gypsum reached their maximum strength at axial

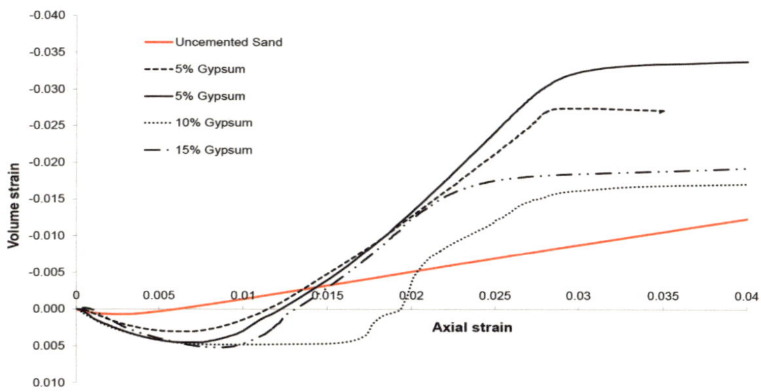

Figure 13.
Volume strain, axial strain responses from drained tests (p'_c = 50 kPa).

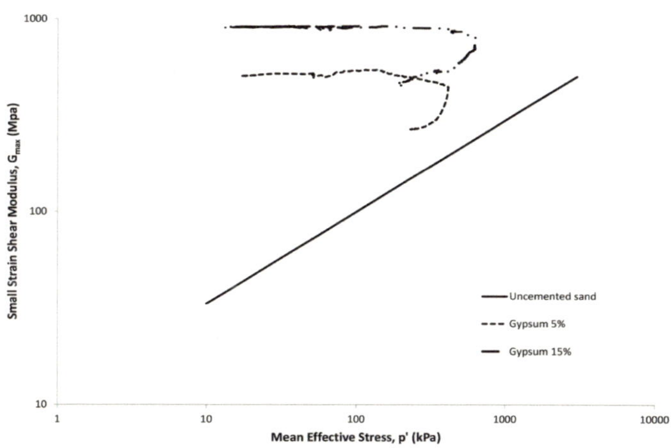

Figure 14.
Variation of G_{max} during compression and shear for gypsum-cemented specimens.

strains <1.5%, while for the uncemented sand, the maximum strength occurred at axial strains of between 2 and 5%. **Figure 13** shows the cementation initially prevents the expansion that occurs almost from the beginning of shear for the uncemented samples. The uncemented samples expand steadily during shear eventually reaching a maximum volume strain of about −0.04 for axial deformations greater than 10%. Even though their densities are similar, the gypsum cemented shows different behavior. The specimens initially compress due to increasing mean stress, but as they approach the peak, they begin to expand at a rapid rate, much more rapidly than the uncemented sand. The rate of expansion then drops as pronounced shear bands develop. In general, these results are typical of the behavior of artificially cemented specimens prepared with a range of cement types [18–37].

Figure 14 shows the response and changes of G_{max} with p′ for typical gypsum-cemented samples and comparison with the response for uncemented sand. The responses include an initial isotropic compression stage to 50 kPa followed by drained shearing to large deformations. The figure shows the cement has a remarkable contribution on the small strain stiffness, with G_{max} nearly constant until reaching the peak strength. However, looking in detail, it is found that G_{max} increases slightly with p′ and then decreases as the cementation begins to break. After the peak, the shear modulus falls significantly and approaches the uncemented response.

6.3 Bio-cemented sand specimens

During the preparation of the UCS and triaxial samples, dry sand was mixed with equal amounts by mass, of powdered calcium chloride and urea, and then with water containing the bacteria (see **Figure 15**). The amounts of calcite precipitated, measured after the mechanical tests, are shown in **Figure 16**. For UCS specimens, there is a clear relationship between the amount of urea added and the amount of calcite. Triaxial samples, on the other hand, show considerable dispersion and variability in the precipitated amount of calcite quantified using acid-wash test. Nevertheless, when the uniformity of calcite precipitation is verified after the test, there is a variability of less than 5% of the amount of calcite in each sample. The calcite concentration in the UCS samples and some typical data were included in **Table 2**.

Figure 16 shows the amount of calcite. However, the amount of calcite in the triaxial specimens has increased. In addition, the variability of the data could be the result of a change of procedure during saturation. In some initial tests, calcite was removed from the samples when pumping water, which could explain some of the lower results. However, in the majority of tests, it is simply a question of pumping water into the samples, and there is no need to lose nutrients or bacteria. The variability of the amount of calcite obtained from the UCS and triaxial tests can be motivated by several reasons. In addition to the hardening time and access to the air causing the drying of the samples, it was the same during triaxial tests; no particular action was taken. Differences in soil temperature and pH during sample preparation may also have contributed to differences in calcite precipitation.

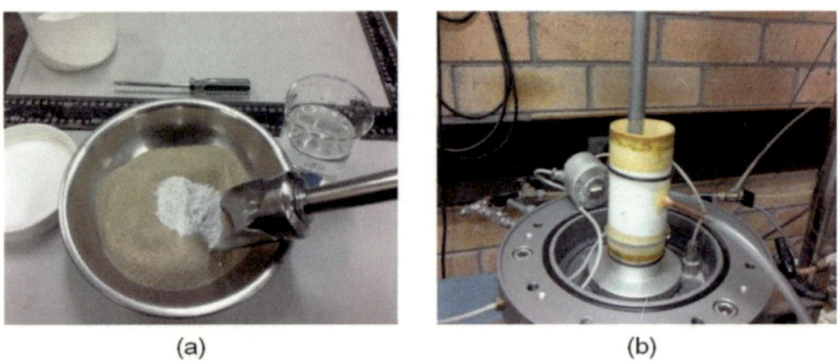

Figure 15.
Procedure of (a) mixing and (b) molding bio-cemented samples for triaxial test.

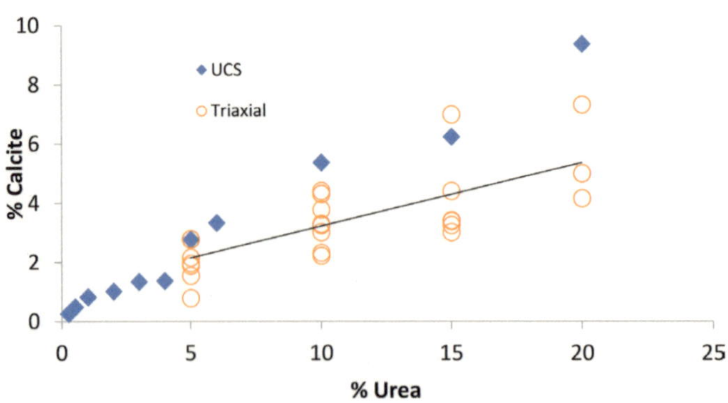

Figure 16.
Relation between amount of urea in mixture and calcite measured posttest.

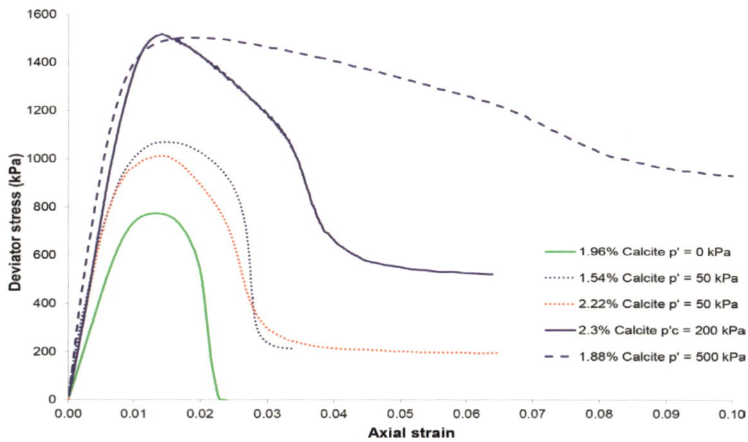

Figure 17.
Deviator stress, axial strain responses for bio-cemented specimens (1.5–2.3% calcite).

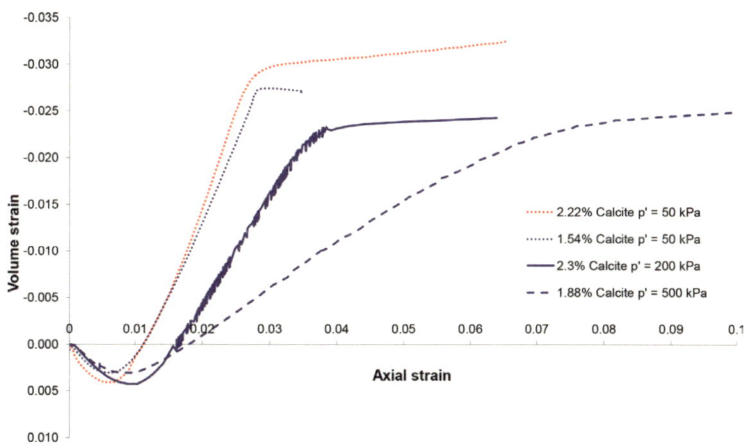

Figure 18.
Volume strain, axial strain responses for bio-cemented specimens (1.5–2.3% calcite).

To study the effects of confining pressure and calcite, different ranges of calcite sample were prepared. In all cases, these differences were prepared in the same way. **Figures 17** and **18** show the stresses, deformations and volume stresses, and axial strains, resulting from a series of CID drained tests with different confinement constraints for the lowest calcite contents, ranging from 1.5 to 2.3%.

The set of tests includes a sample for which the membrane leaks because the cell and the back pressures were equal. This has been done effectively on a totally saturated sample. The UCS tests are shown in **Figure 4**. **Figure 4** suggests a UCS strength of about 300 kPa for a calcite content of 2%. Past research [17] claims that the strength of the UCS in bio-cemented sample is influenced by the level of saturation. However, [17] reported that the degree of saturation causes an increase in UCS, which is the opposite trend of the current study. Previous research [13] has therefore focused on the localization of calcite, with lower degrees of saturation leading to precipitation only at particle contact. In this chapter, all samples have been prepared so that there is no significant saturation effect on the results.

The results shown in **Figures 16** and **17** indicate a general tendency toward strength and stiffness increase with the confining constraint. However, at a

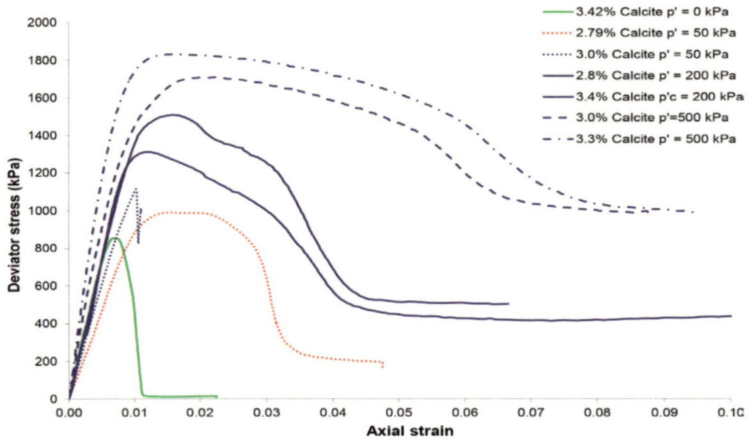

Figure 19.
Deviator stress, axial strain responses for bio-cemented specimens (2.8–3.4% calcite).

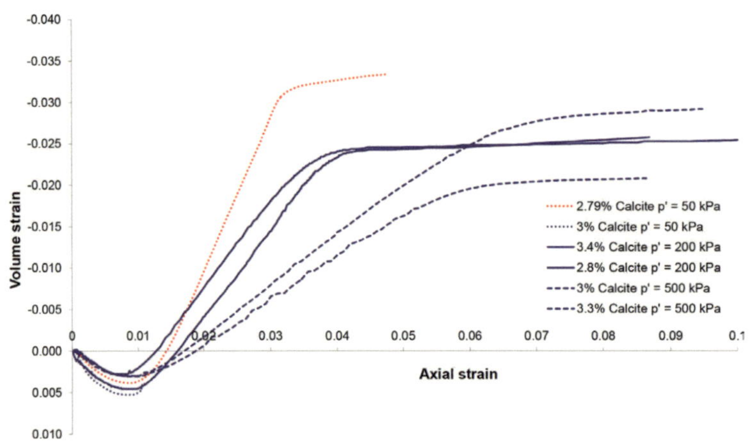

Figure 20.
Volume strain, axial strain responses for bio-cemented specimens (2.8–3.4% calcite).

confining stress of more than 200 kPa, the resistance becomes more obvious. This could be due to the lower calcite content in the more heavily stressed sample. It can also be noted that the cumulative response of the sample subjected to higher stresses shows less compression and more gradual expansion, which corresponds to a lesser effect of cementation. Thus, the calcite content is not only low but also the level of increased stress. Nevertheless, the general behavior patterns correspond to those expected for cemented specimens and are similar to gypsum cement.

Figures 19 and **20** show the effects of confining stress for a series of triaxial CID tests with calcite contents between 2.8 and 3.4%. Another UCS test is available for a saturated test in this cement content range, as previously following the rupture of the membrane. The UCS resistance of 820 kPa is again significantly higher than expected in the UCS tests of **Figure 4**, giving a value of 450 kPa for a calcite content of 3.4%. Reasonably consistent, all bio-cemented specimens showing increased in strength and stiffness as containment stress increases. For lower calcite levels, the rate of expansion tends to decrease as the level of stress increases, although the effect is less pronounced for those more cemented specimens. The trends are generally similar to those of the lower calcite content.

7. Conclusion

The following concluding remarks are made based on the performance and behavior of bio-cemented Sydney sand:

- Bio-cemented samples were prepared by mixing sand, bacteria, and nutrients, as well as samples cemented with gypsum. It has been found that it produces no damage when produced by the sample preparation method. A mixing technique is recommended to study the response of a weakly cemented material. However, there are limits to mixing with the content of the mixture.

- As shown in several other studies, calcite is an extremely effective cementing agent, and, for a given amount of cement, it offers higher strength and stiffness than other cementing agents. The results show that the strength in the UCS tests is similar to, or slightly higher than, the samples treated with injection techniques. At the same time, the problem of injection site obstruction was avoided by using an ex situ mixing technique, and this has been successfully demonstrated as feasible at the laboratory scale.

- The patterns of behavior observed in bio-cemented Sydney sand in triaxial tests are very similar to those of gypsum-related specimens. The results were reasonably consistent throughout the laboratory tests. The results of the triaxial tests were obtained with the amount of calcite produced, and it is difficult to predict the degree of cementation.

- The use of automated shear wave velocity measurement has enabled variations in stiffness, and hence degree of cementation, to be monitored throughout the processes of curing, stress application, and shearing. However, the large changes in shear wave velocity associated with curing have caused some difficulties in obtaining reliable data.

Acknowledgements

This research was conducted in the best interest of the corresponding author's PhD research work. His entire studies were fully sponsored by the Ministry of Higher Education of Malaysia. All the research facilities and the instruments used during the research belong to the University of Sydney.

Author details

Youventharan Duraisamy[1*] and David Airey[2]

1 Department of Civil Engineering, Universiti Malaysia Pahang, Kuantan, Malaysia

2 School of Civil Engineering, The University of Sydney, NSW, Australia

*Address all correspondence to: youventharan@ump.edu.my

References

[1] Jiang NJ, Soga K. Erosional behavior of gravel-sand mixtures stabilized by microbially induced calcite precipitation (MICP). Soils and Foundations. 2019;**59**(3):699-709

[2] Duraisamy Y. Strength and stiffness improvement of bio-cemented Sydney sand[Ph.D thesis]. The University of Sydney; 2016

[3] Chittoori B, Rahman T, Burbank M, Moghal AAB. Evaluating shallow mixing protocols as application methods for microbial induced calcite precipitation targeting expansive soil treatment. In: Geo-Congress 2019: Soil Improvement. ASCE. 2019. pp. 250-259

[4] DeJong JT, Fritzges MB, Nusslein K. Microbially induced cementation to control sand response to undrained shear. Journal of Geotechnical and Geoenvironmental Engineering. 2006;**132**(11):1381-1392

[5] Ivanov V, Chu J. Applications of microorganisms to geotechnical engineering for bioclogging and biocementation of soil in situ. Reviews in Environmental Science and Biotechnology. 2008;**7**(2):139-153

[6] Whiffin VS, Van Paassen LA, Harkes MP. Microbial carbonate precipitation as a soil improvement technique. Geomicrobiology Journal. 2007;**24**(5):417-423

[7] Terzis D, Laloui L. A decade of progress and turning points in the understanding of bio-improved soils: A review. Geomechanics for Energy and the Environment. 2019;**19**:100116

[8] Van Paassen LA, Ghose R, Van Der Linden TJM, Van Der Star WRL, Van Loosdrecht MCM. Quantifying biomediated ground improvement by ureolysis: Large-scale biogrout experiment. Journal of Geotechnical and Geoenvironmental Engineering. 2010;**136**(12):1721-1728

[9] Stocks-Fisher S, Galinat JK, Bang SS. Microbiological precipitation of $CaCO_3$. Soil Biology and Biochemistry. 1999;**31**(11):1563-1571

[10] Bachmeier KL, Williams AE, Warmington JR, Bang SS. Urease activity in microbiologically-induced calcite precipitation. Journal of Biotechnology. 2002;**93**(2):171-181

[11] Dick J, De Windt W, De Graef B, Saveyn H, Van der Meeren P, De Belie N, et al. Bio-deposition of a calcium carbonate layer on degraded limestone by *Bacillus* species. Biodegradation. 2006;**17**(4):357-367

[12] Achal V, Mukherjee A, Basu P, Reddy M. Strain improvement of *Sporosarcina pasteurii* for enhanced urease and calcite production. Journal of Industrial Microbiology & Biotechnology. 2009;**36**(7):981-988

[13] Van Paassen LA. Bio-mediated ground improvement: From laboratory experiment to pilot applications. In Geo-Frontiers 2011: Advances in Geotechnical Engineering. Vol. 397. ASCE; 2011. pp. 4099-4108

[14] Huang J. The effects of density and cementation of soils. [Ph.D thesis]. University of Sydney; 1994

[15] Consoli NC. Fundamental parameters for the stiffness and strength control of artificially cemented sand. Journal of Geotechnical and Geoenvironmental Engineering. 2009;**135**(9):1347

[16] Clough G, Iwabuchi J, Rad N, Kuppusamy T. Influence of cementation on liquefaction of sands. Journal of Geotechnical Engineering. 1989;**115**(8):1102-1117

[17] Cheng L, Cord-Ruwisch R, Shahin MA. Cementation of sand soil by microbially induced calcite precipitation at various degrees of saturation. Canadian Geotechnical Journal. 2013;**50**(1):81-90

[18] Clough GW, Rad NS, Bachus RC, Sitar N. Cemented sands under static loading. Journal of the Geotechnical Engineering Division. 1981;**107**(6):799-817

[19] Duraisamy Y, Airey D. Small strain modulus of bio-cemented sand. In: Deformation Characteristics of Geomaterials: Proceedings of the 6th International Symposium on Deformation Characteristics of Geomaterials, IS-Buenos Aires 2015, 15-18 November 2015. Vol. 6. Buenos Aires, Argentina: IOS Press; 2015. p. 283

[20] Al Thawadi SM. High strength in-situ biocementation of soil by calcite precipitating locally isolated ureolytic bacteria[Ph.D thesis]. Australia: Murdoch University; 2008. p. 264

[21] Van Paassen LA. Biogrout: Ground improvement by microbially induced carbonate precipitation[Ph.D thesis]. Delft: Department of Geotechnology, Delft University of Technology; 2009

[22] Mahawish A, Bouazza A, Gates WP. Unconfined compressive strength and visualization of the microstructure of coarse sand subjected to different biocementation levels. Journal of Geotechnical and Geoenvironmental Engineering. 2019;**145**(8):04019033

[23] Nafisi A, Khoubani A, Montoya BM, Evans MT. The effect of grain size and shape on mechanical behavior of MICP sand I: Experimental study. In: Proceedings of the 11th National Conf. in Earthquake Eng., Earthquake Eng. Research Ins. Los Angeles; 2018. (In review)

[24] Al Qabany A, Soga K. Effect of chemical treatment used in MICP on engineering properties of cemented soils. Geotechnique. 2013;**63**(4):331-339

[25] Choi SG, Hoang T, Park SS. Undrained behavior of microbially induced calcite precipitated sand with polyvinyl alcohol fiber. Applied Sciences. 2019;**9**(6):1214

[26] Gowthaman S, Mitsuyama S, Nakashima K, Komatsu M, Kawasaki S. Biogeotechnical approach for slope soil stabilization using locally isolated bacteria and inexpensive low-grade chemicals: A feasibility study on Hokkaido expressway soil, Japan. Soils and Foundations. 2019;**59**:484-499

[27] Cheng L, Cord-Ruwisch R. In situ soil cementation with ureolytic bacteria by surface percolation. Ecological Engineering. 2012;**42**(0):64-72

[28] Airey D, Mohsin A. Evaluation of shear wave velocity from bender elements using cross-correlation. Geotechnical Testing Journal ASTM. 2013;**36**(4):506-514

[29] Duraisamy Y, Airey DW. Performance of biocemented Sydney sand using ex situ mixing technique. DFI Journal-The Journal of the Deep Foundations Institute. 2015;**9**(1):48-56

[30] Bolton M. The strength and dilatancy of sands. Geotechnique. 1986;**36**(1):65-78

[31] Shibuya S, Tanaka H. Estimate of elastic shear modulus in Holocene soil deposits. Soils and Foundations. 1996;**36**(4):45-55

[32] Kokusho T. Cyclic triaxial test of dynamic soil properties for wide strain range. Soils and Foundations. 1980;**20**(2):45-60

[33] Hardin BO, Richart F Jr. Elastic wave velocities in granular soils. Journal of

the Soil Mechanics and Foundations Division. 1963;**89**:33-65. (Proceeding Paper 3407)

[34] Huang JT, Airey DW. Properties of artificially cemented carbonate sand. Journal of Geotechnical and Geoenvironmental Engineering. 1998;**124**(6):492-499

[35] Sharma R, Baxter C, Jander M. Relationship between shear wave velocity and stresses at failure for weakly cemented sands during drained triaxial. Soils and Foundations. 2011;**51**(4):761-771

[36] Lee M-J, Choi S-K, Lee W. Shear strength of artificially cemented sands. Marine Georesources and Geotechnology. 2009;**27**(3):201-216

[37] Asghari E, Toll DG, Haeri SM. Triaxial behavior of a cemented gravelly sand, Tehran Alluvium. Geotechnical and Geological Engineering. 2003;**21**(1):1-28